U0179820

东南大学建筑学院国际联合教学丛书
International Joint Teaching Series of SEU-ARCH

RURAL REGENERATION

Sino-Italian Joint Teaching Studio on Conservation and Revitalization of Rural Architectural Heritage Liangheko u in Xuan´en

乡村再生

中意联合乡村建筑遗产活态更新研究与教学 · 宣恩两河口

张彤 王川 李保峰 汤诗旷 Enrico Fontanari Aldo Aymonino Giuseppe Caldarola
褚冬竹 覃琳 宫聪 李海清 徐涵 著

东南大学出版社
· 南京 ·

图书在版编目（CIP）数据

乡村再生：中意联合乡村建筑遗产活态更新研究与教学·宣恩两河口 / 张彤等著 . -- 南京：东南大学出版社，2024.3
ISBN 978-7-5766-0607-2

Ⅰ.①乡… Ⅱ.①张… Ⅲ.①乡村规划－研究－中国 Ⅳ.①TU982.29

中国版本图书馆 CIP 数据核字（2022）第 249066 号

责任编辑：戴　丽　魏晓平
责任校对：张万莹
封面设计：王　川
责任印制：周荣虎

乡村再生：中意联合乡村建筑遗产活态更新研究与教学·宣恩两河口
Xiangcun Zaisheng : Zhongyi Lianhe Xiangcun Jianzhu Yichan Huotai Gengxin Yanjiu Yu Jiaoxue · Xuan'en Lianghekou

出版发行：东南大学出版社
出 版 人：白云飞
社　　址：南京市四牌楼 2 号
网　　址：http://www.seupress.com
邮　　箱：press@seupress.com
邮　　编：210096
电　　话：025-83793330
经　　销：全国各地新华书店
印　　刷：南京新世纪联盟印务有限公司
开　　本：889 mm × 1 194 mm　1/20
印　　张：9.5
字　　数：579 千
版　　次：2024 年 3 月第 1 版
印　　次：2024 年 3 月第 1 次印刷
书　　号：ISBN 978-7-5766-0607-2
定　　价：98.00 元

本书内容呈现于

第 17 届威尼斯国际建筑双年展官方平行展

两河口，一个土家会聚之地的再生

Collateral Event of the 17th International Architecture Exhibition-La Biennale di Venezia

LIANGHEKOU, A Tujia Village of Re-Living-Together

国家社会科学基金“铸牢中华民族共同体意识”研究专项
“中华民族共同体视觉形象聚类分析与图谱建构”（20VMZ008）资助

Organizers

Southeast University（SEU）

Huazhong University of Science and Technology（HUST）

Università Iuav di Venezia（IUAV）

Chongqing University（CQU）

Course Supporters and Sponsors

Municipal Government of Xuan´en County

Fondazione EMGdotART

Education Foundation of Southeast University

Tutors

SEU: Zhang Tong, Wang Chuan

HUST: Li Baofeng, Tang Shikuang, Wang Tong

IUAV: Aldo Aymonino, Enrico Fontanari, Giuseppe Caldarola

CQU: Chu Dongzhu, Gong Cong

Participating Students

SEU: Xu Han, Sui Mingming, Chen Siyu, Ma Yumeng, Chen Yunxuan, Shen Jie, Yan Hongyan, Ge Shizhao, Xu Shuang, Wang Rong, Li Xinran

HUST: Zhao Rui, Wei Di, Zhang Shiwei, Wang Peize, Peng Xinyi, Wang Ning, He Shixuan, Tian Shuying, Liu Wenyu (Hubei Institute of Fine Art), Wang Changxi (Hubei Institute of Fine Art)

IUAV: Giovanna Bordin, Ilaria Bottino, Davide Burro, Davide Bruneri, Alberto Canton, Elisa Cielo, Marco D´Altoe, Anna Marsella, Thomas Ortolan, Giacomo Rettore, Ilaria Rosolen, Alessandro Sparapan, Allegra Zen

CQU: Gu Mingrui, Yang Rui, Li Chen, Wu Shuang, Huang Jinjing, Dai Lianjie

组织机构

东南大学
华中科技大学
威尼斯建筑大学
重庆大学

课程支持与资助

宣恩县人民政府
意大利雅伦格文化艺术基金会
东南大学教育基金会

指导教师

东南大学：张彤、王川
华中科技大学：李保峰、汤诗旷、王通
威尼斯建筑大学：阿尔多・艾莫尼诺、恩里科・丰塔纳里、朱塞佩・卡尔达罗拉
重庆大学：褚冬竹、宫聪

参与学生

东南大学
徐涵、隋明明、陈斯予、马雨萌、陈韵玄、沈洁、闫宏燕、戈世钊、续爽、王瑢、李心然
华中科技大学
赵蕊、魏迪、张师维、王沛泽、彭欣怡、王宁、何仕轩、田淑颖、刘文玉（湖北美术学院）、王长曦（湖北美术学院）
威尼斯建筑大学
乔凡娜・博尔丁、伊拉里亚・波提诺、达维德・布诺、达维德・布鲁内里、阿尔伯特・坎顿、埃莉萨・切诺、马克・达尔托、安娜・马尔赛拉、托马斯・奥尔托兰、贾科莫・雷托雷、伊拉里亚・洛塞伦、亚里山德罗・斯帕拉潘、阿莱格拉・泽恩
重庆大学
顾明睿、阳蕊、李晨、吴霜、黄金静、戴连婕

前 言

两河口，位于湖北省恩施土家族苗族自治州宣恩县沙道沟镇，地属武陵山腹地、酉水河上游，是目前武陵山少数民族地区保存最为完好的传统市集聚落之一。由于水陆交通便利，古往今来这里都是商贸集聚之地，也曾是“川盐济楚”的重要枢纽。

近30年来，随着商贸运输方式的转变，尤其是城镇化进程的加快，两河口村的商贸地位逐渐下降，人口外流，曾经兴盛的老街日趋凋敝，仅有保存较好的百米古街依然勾起人们对往昔历史的记忆。近年来，当地政府重视民族文化和地方传统的发掘、整理，以毗邻的国家级重点文物保护村寨彭家寨为代表，在两河口村所在的盐道山谷整体打造土家族泛博物馆，传承和发扬民族优秀文化，推进乡村振兴。两河口老街迎来了应对乡村旅游经济问题、吸引村民返乡创业、重振商业服务功能的良机。

以国内外知名建筑院校国际四校联合研学营为纽带，带动两河口活化更新的动议得到州县两级政府的大力支持。2019年4月至5月，在雅伦格文化艺术基金会的资助下，东南大学、华中科技大学、重庆大学和威尼斯建筑大学的40多名师生进驻两河口村，讨论如何在即将到来的旅游大潮中，避免浅表性消费，保护和传承国家级文化遗产的价值内核，使得沉睡的老街和周边村寨得以振兴再生。这不仅是中国实施乡村振兴战略普遍遇到的问题，也是包括威尼斯在内的人类文化遗产共同面临的挑战。

在为期两个月的课程中，四校师生就两河口老街的活态保护与适应性更新各自开展规划设计和策略研究。2019年6月，四校师生齐聚威尼斯，在威尼斯建筑大学组织设计评图和成果交流，并在雅伦格文化艺术基金会禅宫举办展览。国际四校联合研学营的合作研究和教学得到国际多所知名高校和威尼斯当地的充分关注。2019年12月国际四校联合研学营的深化成果以“两河口，一个土家会聚之地的再生”为题受邀参加第17届威尼斯国际建筑双年展，成为这个久负盛名的国际展会该届15个官方平行展之一。

以两河口老街为对象的建筑遗产保护与更新，在国际比较的语境中展现出一种与西方石构建筑定义的独一性和原真性不同的理念和方法。由于干栏木构建造体系高度发达的装配化、灵活性与可变性，其自身在时间的进程中具备一种自适应、拓扑性的活态更新机制。两河口国际四校联合研学营的研究和实践，意在求证即便当代城乡生活发生了如此巨大激变，这种独具东方智慧的传统依然具备如生命般的自适应更新和可持续再生能力。

文化传承是根脉的延续。两河口老街再生的研究和实践，是基于对以土家族为主的当地多民族文化的珍视、发掘、保护和发扬；同时思考面临旅游业的发展，文化的原真价值如何得到保护和传续，而不是仅仅被消费。在乡村振兴和传统文化复兴成为国家战略的今天，国际四校联合研学营在两河口的探索展现出一条两者协同耦合的路径，文化是乡村振兴的内生资源和驱动力，乡村振兴推动地方文化在更广阔的范围得到理解、尊重和光扬。四校联合研学营的成果受邀参加威尼斯国际建筑双年展，说明以文化价值驱动社会经济发展具有普遍的价值认同。希望在各界共同努力下，以国际四校联合研学营为启端的两河口老街更新实践能为乡村振兴和文化复兴提供一个具有持久价值的中国案例。

2022 年 11 月

INTRODUCTION

Lianghekou, located in Shadaogou Town, Xuan'en County, Enshi Tujia and Miao Autonomous Prefecture, Hubei Province, belongs to the hinterland of Wuling Mountains and the upper reaches of Youshui River. It is one of the best-preserved traditional market settlements in the Wuling Mountains minority area. Thanks to the convenient location for land and water transportation, this village had been a trade center since ancient times and acted as the hub of the "Chuanyan Ji Chu i.e. utilizing the salt of Sichuan to purvey Chu" movement.

In the past 30 years, Lianghekou Village's commercial importance has gradually declined and the population has flowed out, due to the changes of commercial transportation, especially with the acceleration of urbanization. The once flourished ancient street is now falling. Only the well-preserved hundred-meter-long historic street still evokes people's memories of the past history. In recent years, the local government has attached great importance to the excavation and arrangement of ethnic culture and local traditions. Represented by the adjacent village of Pengjia Village, a national-level key cultural relic protection village, a pan-museum of the Tujia nationality has been built in the Salt Road Valley where Lianghekou Village is located. It promotes the excellent ethnic culture and boosts the village's revitalization. Lianghekou old street now has the ideal opportunity to respond to rural tourism economy issues, attract villagers to move back and start their own businesses, and revitalize the commercial service.

The motion of the international joint studio of world-renowned architectural colleges was strongly backed by the local governments of Enshi Prefecture and Xuan'en County. From April to May 2019, with the support of the Fondazione EMGdotART, more than 40 teachers and students from Southeast University, Huazhong University of Science and Technology, Chongqing University and Università Iuav di Venezia settled in Lianghekou Village to discuss how to avoid superficial consumptions, preserve and pass down the national-level cultural heritages' core value, and revivify the sleeping ancient village amid the approaching tourism upsurge, which is not only a common question for China's actions to its ambitious rural revitalization strategy, but also the challenge faced by the cultural heritage of mankind, including Venice.

During the two-month design studio, teachers and students from the four universities carried out their own planning and strategy research on the flexible protection and adaptive renewal of Lianghekou historic street. In June 2019, teachers and students gathered in Venice to evaluate and exchange their research and design outcomes at Università Iuav di Venezia and held an exhibition in the Palazzo Zen of Fondazione EMGdotART. The joint research and teaching activity of the international joint studio received wide attention from Venice and universities across the world. In December 2019, the further developed outcomes of the international joint studio, titled "Lianghekou, a Tujia Village of Re-Living-Together", was invited to participate in the prestigious 17th International Architecture Exhibition—La Biennale di Venezia as one of its 15 official collateral events.

In the context of international comparison, the architectural heritage protection and renewal of Lianghekou old street demonstrates a concept and method different from the homogenization and authenticity of Western stone-based architecture. Ganlan wooden structure has strong adaptive and topological renewal features in the flow of time as it has a highly developed prefabrication system which is exceptionally flexible and variable. Lianghekou international joint studio and its subsequent events prove that even though contemporary urban and rural life has undergone such drastic changes, this tradition with unique oriental wisdom still has the ability of self-adaptive renewal and sustainable regeneration.

Cultural inheritance is the continuation of roots. The research and practice of the regeneration of Lianghekou old street is based on cherishing, excavation, protection, and promotion of the local multi-ethnic culture represented by the Tujia nationality, as well as protecting and continuing local culture's authenticity value rather than being consumed during the tourism development. Today, when rural revitalization and traditional culture regeneration become national strategies, the actions of the international joint studio in Lianghekou have found a path to combine these two. Culture is the endogenous resource and driving force of rural revitalization, while rural revitalization supports local culture to be better understood, respected and spread in a broader range. The achievements of the international joint studio, particularly the invitation to participate in the Venice Biennale of Architecture, show that cultural value to drive social and economic development has universal value recognition. We hope that, with collaborative efforts of different sectors, the renewal of Lianghekou old street initiated by the international joint studio can provide a Chinese case with lasting value for rural revitalization and cultural revival.

ZHANG Tong
Nov. 2022

纪念亲爱的恩里科·丰塔纳里教授，

怀着感激与美好的回忆。

In memory of dear Professor Enrico Fontanari,
with gratitude and fond memories.

目录

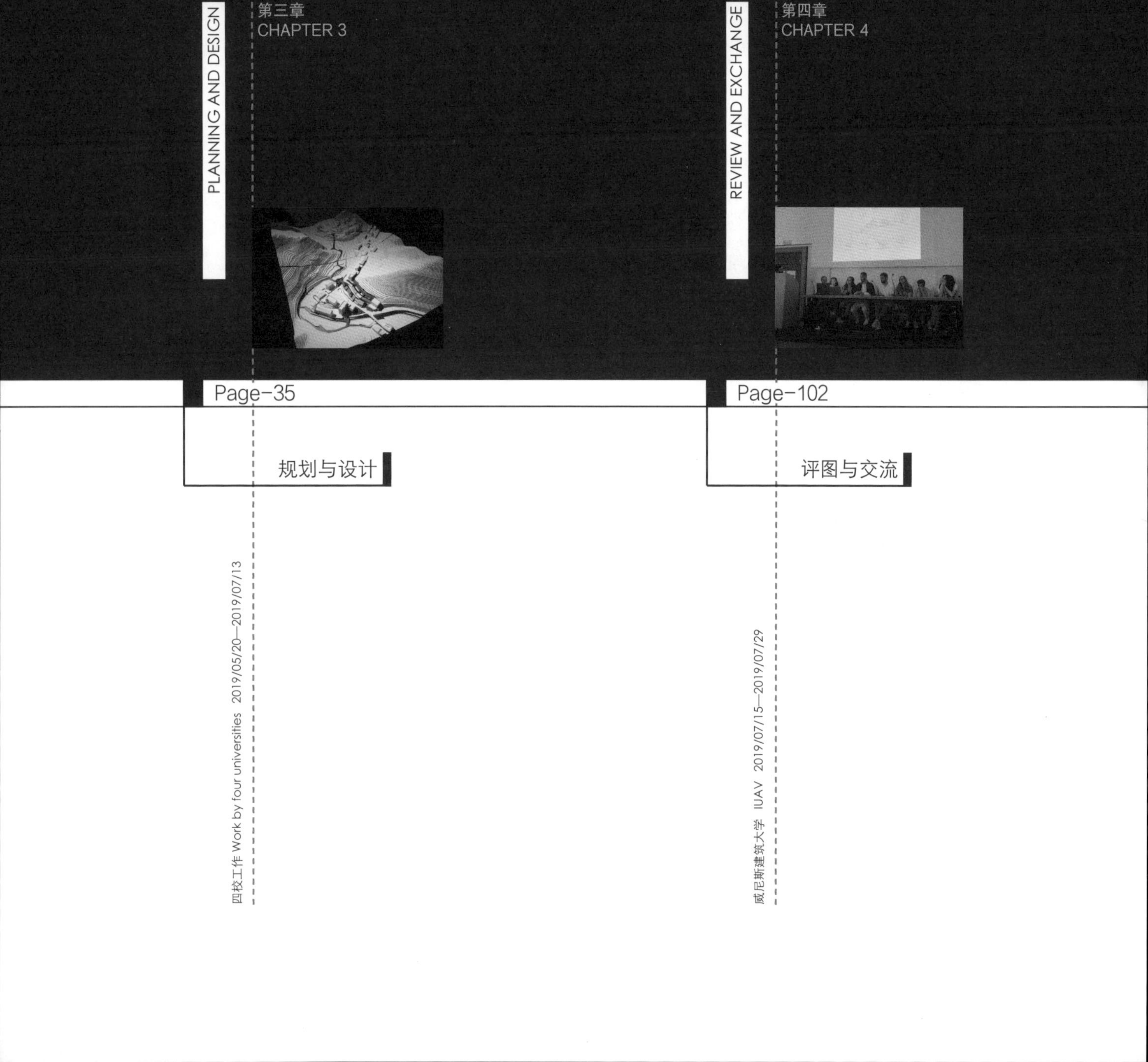

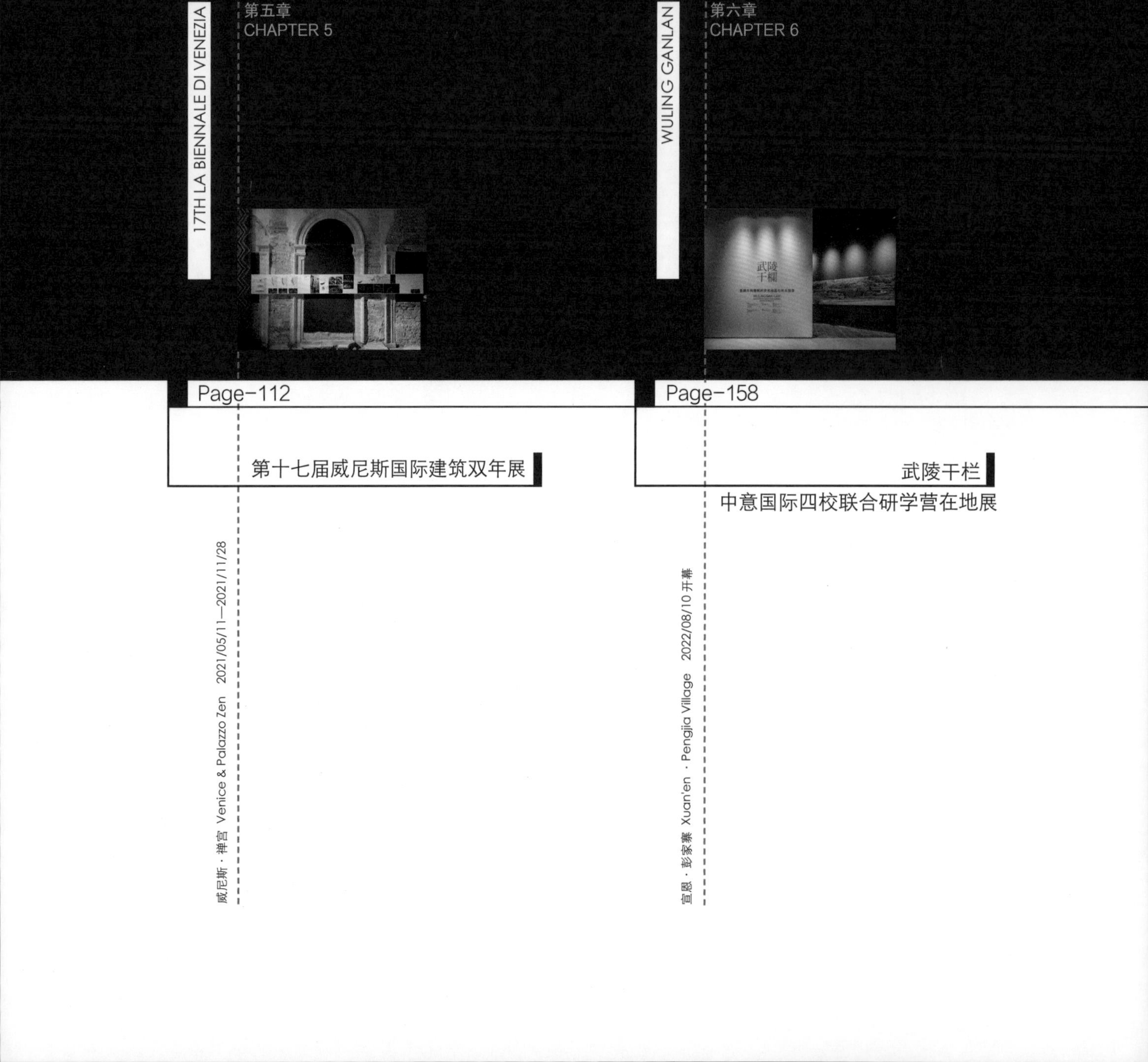

第七章
CHAPTER 7

POSTCRIPT AND ACKNOWLEDGEMENTS

后记与致谢

南京 Nanjing 2022/11/19

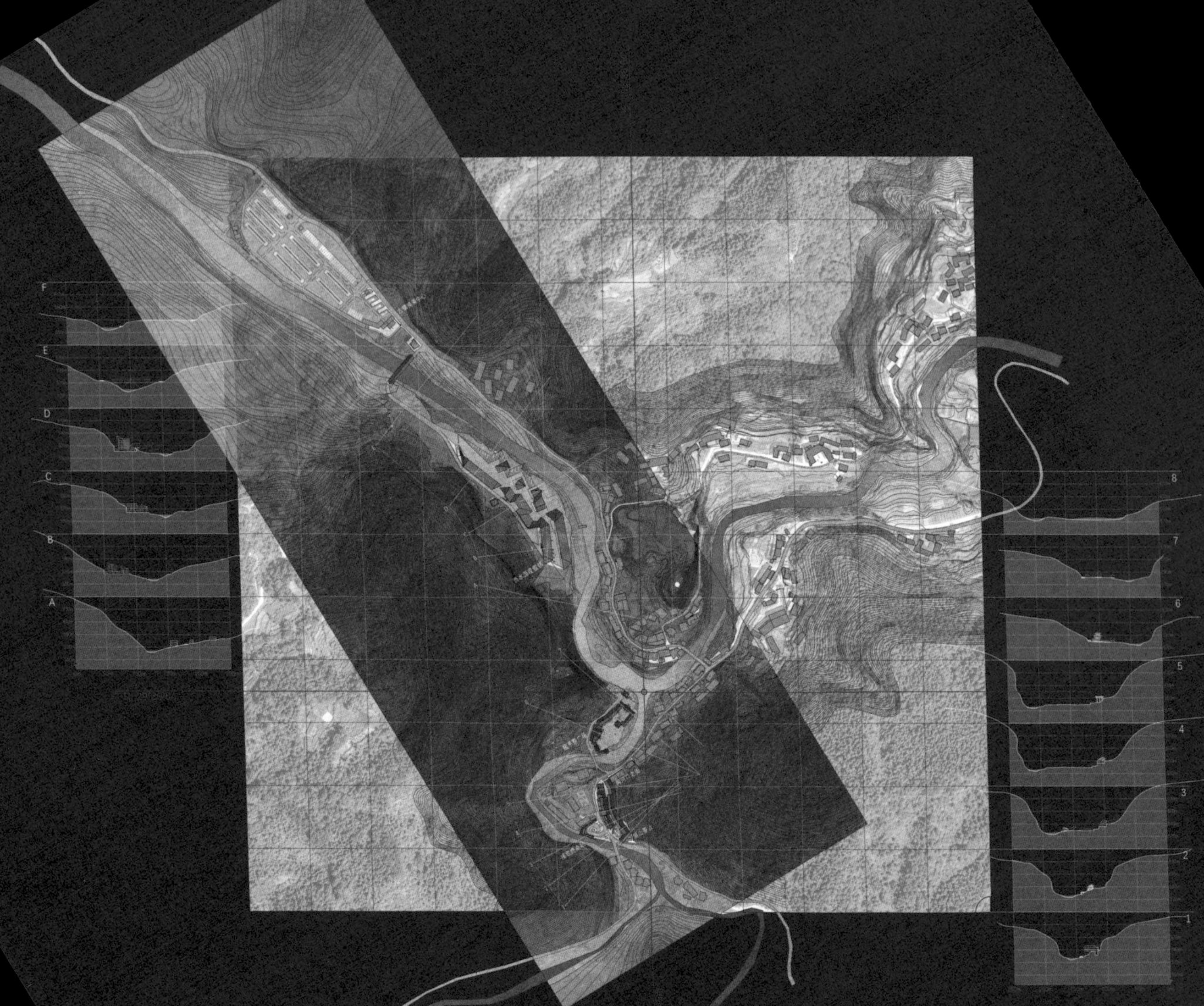
F
E
D
C
B
A
8
7
6
5
4
3
2
1

教学与研讨 TEACHING AND DISCUSSION

武陵土家乡土建筑田野调查
与两河口村现场团队测绘工作
Field investigation of Tujia rural architecture in the Wuling mountains area
& teamwork of measuring surveys of Lianghekou Village
27-Apr-2019 to 5-May-2019

四校的两河口村规划建筑方案设计
Planning and architectural schemes of Lianghekou Village by four universities
27-Apr-2019 to 5-May-2019

东南大学 SEU
木构变色龙：从日常生活到节日庆典
Wooden Chameleon: From Daily Life to Festival
土家民艺村：公共空间和传统吊脚楼活化
Tujia Artware Village: Regeneration of Public Space and Traditional Stilted buildings
土家泛文化民宿村设计
Tujia Pan-culture Homestay Village Design

华中科技大学 HUST
两河集事
Story Market
言（盐）·趣：两河口盐道古街复兴
Dialogue · Interest: Regeneration of the Salt Road Ancient Street in Lianghekou

重庆大学 CQU
乡村生命体：历史 | 健康 | 和谐
Village Organism: History, Health and Harmony

威尼斯建筑大学 IUAV
竹钢
The Green Steel
重塑土家遗产
Reframing Tujia's Legacy
穿山小径
Path Through the Hillsides
土家门户
TujiaThreshold

野椒园·宣恩壹号粮仓测绘工作 李海清教授团队
Measuring survey of Xuan'en No.1 Granary in Yejiao Garden
by Prof. Li Haiqing' steam
15-Aug-2019 to 28-Aug-2019

评图与交流
Review and exchange
07-Jun-2019 to 20-Jul-2019
威尼斯建筑大学中期评图汇报
Mid-term presentation and review in IUAV
威尼斯禅宫教学成果展览
Exhibition of workshop archievements in Palazzo Zen

威尼斯建筑大学深化学习设计
Further learning and optimize design proposals in IUAV
15-Aug-2019 to 28-Aug-2019

申请第十七届威尼斯国际建筑双年展，并获批策展为官方平行展：
两河口，一个土家会聚之地的再生
Apply for the17th International Architecture Exhibition—La Biennale di Venezia and be approved as an official collateral exhibition: Lianghekou, a Tujia Village of Re-Living-Together
10th-Sep-2019 to 20th-Oct-2019

第十七届威尼斯国际建筑双年展官方平行展：两河口，一个土家会聚之地的再生
A collateral exhibition of 17th International Architecture Exhibition—La Biennale di Venezia: Lianghekou, a Tujia Village of Re-Living-Together
地点：威尼斯禅宫 Location: In Palazzo Zen
21-May-2021 to 28-Nov-2021

亚洲干栏木构建筑遗产的保护与再生学术研讨会
International Symposium on Conservation and Regeneration of Asian Ganlan Wooden Architectural Heritage
10th-Aug-2022 to 11st-Aug-2022
宣恩县两河口村中意国际四校联合研学营
Sino-Italian four universities international joint studio in Lianghekou Village, Xuan'en County

武陵干栏：亚洲木构建筑的文化结晶与样本遗存展览
Wuling Ganlan:
Splendid Archi-tectonic Legacy of Asian Wooden Architecture
地点：宣恩两河口村中意国际四校联合研学营展厅
Location: Exhibiton Hall of Sino-Italian four universities international joint studio in Lianghekou Village, Xuan'en County
10-Aug-2022 to 11-Aug-2022

第十八届威尼斯国际建筑双年展中国国家馆主题展览
更新·共生 Renewal: A Symbiotic Narrative
两河口，一个土家村寨的自适应与拓扑再生
Lianghekou, a Tujia Village of Self-adaptability and Topological Regeneration
意大利威尼斯军械库
Venice Arsenale Magazzino delle Cisterne
21-May-2023 to 28-Nov-2023

交流与展览 EXCHANGE AND EXHIBITION

第一章 历史与文化背景
CHAPTER 1 HISTORY AND CULTURAL CONTEXT

1.1 土家族人

土家族是中国的一个少数民族，主要分布在中国的湘、鄂、渝、黔毗邻的武陵山区：湖南湘西的永顺、龙山、保靖、古丈、凤凰、吉首、泸溪、大庸、桑植、慈利、武陵源等县市区；湖北鄂西的来凤、鹤峰、咸丰、宣恩、恩施、利川、建始、巴东、长阳、五峰等县；重庆东南的黔江区和石柱、酉阳、秀山、彭水等县；贵州的沿河、印江、思南、铜仁、德江、江口等县。

土家族是中国通过民族识别在1957年公开确认的一个单一少数民族，也是中国人口较多的少数民族之一。根据2010年第六次全国人口普查，土家族人口数为8 353 912人，在50多个少数民族人口中排第七位。

土家族自称“毕兹卡”，土家语意为“土生土长的人”。土家族没有本族文字，但是有民族语言，分为南部方言和北部方言。这一语言被认定为属于汉藏语系藏缅语族的一种单一语言。土家族人的族源有多种说法，当代认同较多的是以古代巴人为主的多源族群融合，其祖先的传说主要发生在酉水流域。

土家族地区在历史上很早就有了行政建制。在唐宋时期，中央政府在土家族地区由当地首领实施管理，史称“羁縻政策”。从元代起，土家族地区开始建立军政合一的土司制度，土司、土官由中央王朝任命，拥有一定数量的武装，同时也是各自区域内最大的封建领主。元至清初，中央政府在土家族地区设立了数十个土司。自清雍正五年（1727年）开始，清政府在土家族地区实行“改土归流”政策，到乾隆末年，基本完成。改土归流打破了“汉不入峒，蛮不出境”的禁令，民间往来日益频繁，教育、文化和技术交流广泛，土家人的生产方式和生活方式发生了重大改变。当前中国的土家族聚居地区采用民族自治方式，主要有州、县两个不同行政级别。部分州、县为土家族与苗族共同的自治行政区域。

土家族主要聚居的武陵山区有土家、苗、瑶、侗等多个少数民族。近代的生产生活交流、文化教育交流、商业流通与技术传播等多因素，促成了土家族聚居区土、苗、汉的民族混居情况。1949年以后，从国家到地方展开的民族识别工作，在大量田野调查与实证基础上，对土家族独特的语言、信仰、物质文化等方面进行了鉴别，并在土家族的民族身份被认定后，对其传统文化要素陆续开展了挖掘、保护、传承的工作。

（作者：覃琳）

土家族老人 Tujia elderly
图片来源：http://todayspot.fbs.one

1.1 Tujia People

Tujia is an ethnic minority in China, mainly distributed in Wuling mountain areas adjacent to Hunan, Hubei, Chongqing and Guizhou: Yongshun, Longshan, Baojing, Guzhang, Fenghuang, Jishou, Huxi, Dayong, Sangzhi, Cili and Wulingyuan counties in Western Hunan; Laifeng, Hefeng, Xianfeng, Xuan'en, Enshi, Lichuan, Jianshi, Badong, Changyang and Wufeng counties in Western Hubei; Qianjiang district and Shizhu, Youyang, Xiushan, Pengshui counties in Southeast Chongqing; Yanhe, Yinjiang, Sinan, Tongren, Dejiang, Jiangkou counties in Guizhou.

This ethnic group is publicly recognized by China through ethnic identification in 1957. It is also one of the ethnic groups with a large population in China. According to the sixth national census in 2010, the population of Tujia is 8,353,912, ranking seventh among more than 50 ethnic minorities.

Tujia people call themselves "*bizika*", which means "native people". Tujia has no native language, but it has a spoken language, which can be divided into southern dialect and northern dialect. This language is regarded as an independent language of the Tibetan-Burmese group of the Sino-Tibetan language family. There are many opinions about the origin of Tujia people. The contemporary identity is the integration of multi-source ethnic groups with ancient Ba people as the main group. The legend of their ancestors mainly took place in Youshui River basin.

Tujia area had an administrative system in the very early days. In the Tang and Song Dynasties, the central government was managed by local leaders in Tujia area, which was known as "Jimi policy". Since the Yuan Dynasty, the *Tusi* system of military and political integration was established in the Tujia area. The *Tusi* and officials were appointed by the central government, with a certain number of armed forces. At the same time, they were also the largest feudal lords in their respective regions. From the Yuan Dynasty to the early Qing Dynasty, the central government set up dozens of chieftains in Tujia areas. Since the fifth year of Yongzheng reign of Qing Dynasty, the government carried out the policy of "*Gai-tu-gui-liu*" in Tujia area, and it was basically completed in the end of Qianlong reign. The reform broke the ban of "Han people don't enter the cave and barbarians don't leave the country". Non-governmental exchanges have become more and more frequent in education, culture and technology. The Tujia people's mode of production and way of life have changed greatly. At present, Tujia-inhabited areas in China adopt the mode of autonomy, which mainly includes two different administrative levels: prefecture and county. Some prefectures and counties are the common autonomous administrative regions of Tujia and Miao.

In Wuling mountain area where Tujia mainly live, there are many ethnic minorities such as Tujia, Miao, Yao and Dong. Modern production and life exchanges, cultural and educational exchanges, commercial circulation and technology dissemination and other factors contributed to the mixed living of Tujia, Miao and Han people in Tujia-inhabited areas. After 1949, the work of ethnic identification from the state to the local areas, based on a large number of field surveys and empirical studies, has identified Tujia's unique language, beliefs, material culture and other aspects. After the ethnic identity of Tujia was identified, the traditional cultural elements of Tujia were excavated, protected and inherited.

(Author: Qin Lin)

土家族人 Tujia people
图片来源：http://todayspot.fbs.one

街头的“三棒鼓”表演，形式类似杂技
The "three-stick drum" performance in the street is similar to acrobatics
图片来源：http://www.cncul.org

场镇中的人们在过年前打糍粑
People in towns make Ciba before Chinese New Year
图片来源：http://www.cncul.org

酒文化习俗：喝咂酒
Wine culture and custom: drinking Zajiu
图片来源：http://www.cncul.org

1.2 土家族文化

土家族在民族信仰、建筑艺术、服饰、饮食及生活礼仪等方面有自己的习俗，有大量传统节日。基于传统民俗，土家族在音乐、舞蹈、戏曲等方面保留了丰富的非物质文化遗产。这些文化内容与信仰、生产、生活仪式密切关联，在当代仍然有一些质朴的传承。

在宗教信仰上，土家族以多神崇拜为显著特征，自然崇拜、图腾崇拜、祖先崇拜构成了土家族宗教信仰的基本内容。图腾崇拜反映了土家族对人与动物之间性灵相通的理解，其中尤以白虎崇拜最为典型。祖先崇拜包括氏族祖先、土王和家先崇拜等。土家族的宗教职业人员称“梯玛”，土家语意指“能通神的人”，他们主持祭祀礼仪，也称“土老师”，作为人神沟通的使者，在土家族社群中有着显赫的地位。和某些民族全职从事宗教活动的人不同，“梯玛”只有在受人延请时开展各种宗教活动，平时不脱离生产和世俗生活。

土家族的口头民俗主要通过歌唱方式反映在日常生活、生产、宗教仪式中，如表述日常生活的情歌、竹枝歌，生活生产仪式中的哭嫁歌、丧鼓歌、上梁歌，结合生产活动的薅草锣鼓歌，宗教仪式中的梯玛神歌等。

土家族的风俗民俗事项包括典型的民俗信仰和民间游戏、体育娱乐、节日、舞蹈等方面，很多结合了口头民俗的歌唱内容。它们大致可以分为娱乐类和仪式类，而节庆活动可能同时属于两者。结合民间体育活动的“三棒鼓”是土家族一种广受喜爱的曲艺走唱形式，是一项同时结合打击乐器、口头表演、体育技艺的活动，形式类似杂技，对于表演者有很高的技巧要求。土家族仪式类活动中多辅以民间舞蹈，舞蹈种类繁多，一般以音乐或歌声伴奏，许多舞蹈带有浓郁的巫术和宗教色彩。典型的有配合“摆手”活动的“摆手舞”，祭祀戏剧活动类的“毛古斯舞”，按照“梯玛”在法事活动中的动作编排的祭祀舞蹈“八宝铜铃舞”，模仿猴子的民间娱乐性舞蹈“厄巴舞”，巫、傩文化结合的“跳马”仪式中酬神的“跳马舞”等。

土家族的物质民俗包括建筑艺术、服饰文化、餐饮文化以及其他民间工艺等。土家族的民居建筑属于中国传统木结构中的穿斗结构。山地民居的两侧厢房在进行建造时，往往架空地面，将立柱落于标高不同的山坡地面，形成吊脚楼。土家族吊脚楼有单侧、双侧转角，甚至三侧贯通的外廊，被称为“走马转角”楼。土家族人善于纺织，有特色织锦“西兰卡普”。

在饮食文化上，土家人有自己的酒文化、茶文化、餐桌习俗。土家酒以高粱、谷类、玉米等粮食酿制。土家族人喜欢辛辣食品，有熏制腊肉的习惯。土家族过年时除了有和周边山区相同的“杀年猪”的习俗外，还有用糯米做糖糁、糍粑，推豆腐及提前一天“过赶年”的习俗。糯米糍粑便于保存，腊肉和豆腐利用火塘熏烤后也可以长期存放，这些反映了传统状态下山地民族的生活智慧。

（作者：覃琳）

做西兰卡普的妇女
The woman is making the Silankarp
图片来源：罗彬，辛艺华 . 土家族民间美术 [M]. 武汉：湖北美术出版社，2004.

西兰卡普纹样
Silankarp patterns
图片来源：罗彬，辛艺华 . 土家族民间美术 [M]. 武汉：湖北美术出版社，2004.

1.2 Tujia Culture

Tujia people have their own customs in ethnic beliefs, architectural art, clothing, diet and life etiquette, and a large number of traditional festivals. Moreover, based on the traditional folk customs, they have retained a wealth of intangible cultural heritage in music, dance, opera and other aspects. These cultural contents are closely related to belief, production and life rituals, and still have some simple inheritance in contemporary times.

In terms of religious beliefs, the Tujia people are characterized by polytheism. Nature worship, totem worship and ancestor worship constitute the basic content of Tujia people's religious beliefs. Totem worship reflects Tujia people's understanding of the spiritual connection between humans and animals, especially the white tiger worship. Ancestor worship includes clan ancestors, *Tuwang* and family ancestors worship . The religious professionals of Tujia people are called *"Tima"*, which means "people who can communicate with gods". They preside over sacrificial rites, also known as "*Tu* teachers". As messengers of communication between people and gods, they have a prominent position in the Tujia community. Unlike some people who are engaged in religious activities full-time, Tima only carries out various religious activities when invited, and usually does not break away from production and secular life.

Tujia oral folk customs are mainly reflected in daily life, production and religious ceremonies through singing. For example, love songs and bamboo branches songs describing daily life, crying wedding songs, mourning drum songs and *Shangliang* songs in life and production ceremonies, songs of weeding grass, gongs and drums combined with production activities, and *Tima* God songs in religious ceremonies.

Tujia folk customs include typical folk beliefs and folk games, sports entertainment, festivals and dances, and many of them also combine the singing content of oral folk customs. It can be roughly divided into entertainment and ceremony, and festival activities may include two aspects at the same time. "Three-stick drum" combined with folk sports activities is a popular form of Tujia folk music walking singing. It is an activity form combining percussion instruments, oral performance and sports skills at the same time. The form is similar to acrobatics, which has high skill requirements for performers. Tujia ritual activities are mostly supplemented by folk dance. There are many kinds of dances, usually accompanied by music or songs. Many dances are full of witchcraft and religious color. Typical ones include hand waving dance in conjunction with the "hand waving" activity, *Maogusi* dance in sacrificial drama activities, the sacrificial dance *Babaotongling* dance arranged according to the movements of *Tima* in religious activities, folk entertaining Eba dance which imitates monkeys, and God worshiping horse dance in the horse dance ceremony which combines with Wu and Nuo cultures, etc.

The material folk customs of Tujia people include architectural art, dress culture, catering culture, and other folk crafts. Tujia people's residential buildings belong to the traditional Chinese wooden structure of *Chuandou* structure. During the construction of the wing houses on both sides of the mountain houses, the ground is often elevated, and the columns fall on the hillside ground with different elevations to form a stilted building. The Tujia stilted building has one-side, two-sides corner, even three sides through the veranda external porch, known as *Zouma* corner building. Tujia people are good at weaving and weaving brocade Silankarp.

In terms of food culture, Tujia people have their own wine culture, tea culture and table customs. Tujia liquor is brewed with sorghum, cereal, corn and other grains. Tujia people like spicy food and have the habit of smoked bacon. In addition to the same "pig killing ritual" custom in the surrounding mountainous areas, Tujia people also use glutinous rice to make grits, glutinous rice cake, and tofu, and have the custom of "catching up with the new year" one day in advance. Glutinous rice cake is easy to preserve, and bacon and tofu can be stored for a long time after being smoked and roasted in the fire-pit, which also reflects the wisdom of life of the mountain people in the traditional state.

(Author: Qin Lin)

乡村中的人们在过年前打糍粑
People in villages make glutinous rice cake before Chinese New Year
图片来源：http://todayspot.fbs.one

鄂西土家族聚落 Tujia settlements in Western Hubei
图片来源：作者自摄

1.3 土家聚落与干栏木构

武陵民族走廊东北部区域的湘鄂西地区，是具有悠久历史和地域特色的土家族聚居地。山峦河流间的土家聚落，因所处地理条件和人文环境的独特性而呈现别具一格的形态风貌。

当地有大山、大河，也有浅丘、溪流，土家聚落就是在同山水环境融合适应中生成与发展的。按地理位置，这里的聚落可分成多种类型，如山地村寨、滨水村寨，当然也有背山面水的村寨、背山临田的村寨，还有散点布局的村寨。山地村寨布局大致有两种特征：一种是水平展开的，我们称为“沿等高线”的布局，例如彭家寨，它的房舍基本上沿着几条等高线横向延展。房屋处于不同高差位置，所以从正面看过去有明显的高低错落感，上下层在不同的标高横向展开。这种沿着山脚展开的“带状”布局的村街，可称为“横街”。另一种是沿着上山的主要路径布置房舍，或从水边码头逐级升高，形成阶梯状的聚落格局。聚落从下往上形成具有显著高差、街面向上延伸、或陡峻或缓坡的立体的画面，常被称为“天街”。

滨水类型的寨子在土家族的聚落中也很常见。房子被土家人直接从岸边延伸到水上，用木柱再支撑起来，形成滨水吊脚楼。沿岸的房舍形成顺河走向的街道，常被称为“河街”。

在恩施地区还有一类商贸集镇，为获得更多的展示交易空间，将街道两侧的屋檐尽可能向街内延伸，街道上空形成“一线天”，而街内多处于可遮蔽日晒雨淋的廊棚之下，当地人称之为“雨街”或“凉亭街”。湘鄂西土家聚居区小型的村寨也非常多，三五户人家的村落常常掩映在绿水青山中，属于“散点式”聚落。

无论靠山还是滨水聚落，土家村寨多为一幢幢干栏木构建筑——吊脚楼居住组团集聚而成。在山地聚落为了获得便利于生活的平场，除了在山坡上辟出有限的“平坝”，土家人还非常智慧地筑起一个个适应坡地的木构平台，平台立柱向下落在坡地上成为支撑的“吊脚”，而向上延伸就成为房屋的梁架“排扇”，相邻排扇之间由横枋和檩条连接，上搁置木椽，铺青瓦，周围以板壁围合，这就是山地吊脚楼。而将这种建构方式运用到河边，沿岸建台造屋，就形成了滨水吊脚楼。吊脚楼的形式很多，甚至平地也可以做吊脚楼。这类架空木构平台的建构方式被称为“干栏”。

上：老屋基吊脚楼 Top: Stilted building on the old stone foundation
下：李氏庄园入口内院 Bottom: Inner courtyard entrance of Li´s Mansion
图片来源：作者自摄

干栏作为一种木构形式，具有很强的适应地貌环境的能力。因此，在多雨潮湿的中国南方地区，干栏木构建筑十分普遍。除土家族外，苗、侗、壮、瑶、布依、傣、黎等民族的传统聚居方式都是木构干栏。干栏平台以上是一家人居住的房舍，通常包括落地的堂屋、烧火做饭的火塘间、一头落地另一头吊起的厢房，日常生活的起居、用餐和卧室都在这个平台之上。而平台下还有一些空间，也常常被利用起来，山地吊脚楼底层空间一般用来圈养牲口和家禽。因此才有“人家吊脚楼下有小羊叫”（沈从文《鸭窠围的夜》）的场景。而沿河的吊脚楼平台向河中延伸，立柱吊在河沿上，柱子之间甚至可以成为泊船的小港或人行的通道。沈从文小说中所描绘的吊脚楼，多半是沿河的吊脚楼。

作为干栏建筑的土家吊脚楼，通常根据建造规模分为4种类型:

第一种为吊脚楼基本型，是在一个平整的“坝子”上建一排3~5间的房屋，呈“一”字形，也称为正屋或“座子屋”，有“一明两暗三开间”或“一明四暗五开间”的布局。事实上，座子屋一般少有“吊脚”，除非场地无法平整，才以吊脚支撑，称为“吊一头”。

第二种是在正屋的左或右端，向正面伸出一个垂直于座子屋的横屋，在湘鄂西地区称为“龛子”。正屋+龛子，一正一横，整体形态就像一把老式的钥匙，故称为“钥匙头”。因为坡地地形，钥匙头的楼屋便建成架空或半架空的形式，这样才出现真正的吊脚。

第三种则是在座子屋的两端各连接一个龛子，就是两把钥匙头接起来，形成“U”字形格局，当地人形象地称其为“撮箕口”或“一担挑”。撮箕口其实已经形成了三合院。

第四种就是在“撮箕口”外边增加一道门房倒座，这就形成了四边围合的“天井院”。做到天井院，规格已经比较高了。大户人家还有可能建成天井院群，那就是庄园了，如利川大水井李氏庄园。

这些以干栏木构为特征的不同类型的吊脚楼组合形成了形态极为丰富的土家聚落。土家人祖祖辈辈聚居在这样的村寨中，一直延续至今，成为当今时代民族聚居文化的活着的“标本”。总的来说，土家聚落文化丰富多彩，也独具特色。从适应自然的聚居建构到形成特定的聚落生活形态，既具有真实、完整的珍贵的遗产价值，又保持了独具特色的聚落风貌。

（作者：李晓峰）

1.3 Tujia Settlements and Ganlan Structure

The areas of Hunan and western Hubei in the Northeast of Wuling ethnic corridor are Tujia settlements with strong regional characteristics and a long history. Situated near mountains and rivers, Tujia settlements demonstrate unique styles thanks to its special geographic and cultural features.

Nurtured by the process of fitting into and living with mountains and hillocks as well as rivers and brooks, Tujia settlements can be categorized according to their locations, for example, mountain villages, waterfront villages, villages fronting water and with a hill at the back, villages fronting farmland and with a hill at the back, and scattered villages. There are, broadly speaking, two main types of mountain villages: type one extends horizontally, which is called "along the contour lines". Typical examples include Pengjia Village, houses of which spread out along a series of contour lines. Therefore, it is easy to notice that the village is at different heights from the front as the upper and lower floors of houses expand horizontally at different altitudes. These village streets that fringe the foot of mountain can be referred to as *hengjie* (horizontal streets). Another type is built in a ladder shape along main roads up the mountain or waterfront wharfs vertically. These settlements extend from bottom to top, with village streets spread steeply or gently, to form a multi-level pattern. Village streets like these are often called *tianjie* (sky street).

Waterfront villages are also common, where Tujia people extend their houses from shore to river by building stilted buildings (buildings with hanging feet) supported by wooden pillars above the water. Streets of these villages are known as *hejie* (river streets) as houses stretch along the river flow.

Market towns in Enshi belong to another category: here, houses extend their eaves to the street as far as possible to obtain more space for demonstration and trading. As a result, it forms a *yixiantian* (thin line of sky) scene over the street, and covers most of the street from exposing to sun and rain with canopies. Local people refer to these streets as "*yujie*" (rain streets) or "*liangtingjie*" (pavilion streets). Plenty of small Tujia villages are distributed widely in Hunan and western Hubei. Consisting of three or five families, these small villages are often hidden in the green mountains. These settlements belong to the "scattered type".

Whether facing water or having a hill at the back, Tujia villages generally consist of clusters of Ganlan wooden architectures—stilted buildings. Living in the mountains, Tujia people not only hew out limited "*pingba*" (flatland) on hillsides, but also, very wisely, build wooden platforms on slopes, by using columns that extend down to the ground as the load-bearing hanging feet and up to the sky as beam frames "*paishan*" (row of screen). Adjacent *paishan* are connected by "*hengfang*" (square-shaped crossbars) and purlins.

三角庄“钥匙头”民居
Yaoshitou Residential Building in Sanjiaozhuang
图片来源：作者自摄

Wooden rafters are put on top of *paishan*, on top of which lay grey tiles. The building is then enclosed by wooden partitions—this is a mountain stilted building. When the construction is built by the river along the shore, it is a waterfront stilted building. There are many forms of stilted buildings. It can even be built on the plains. The construction method of building aerial wooden platforms like these is referred to as "Ganlan".

As a wooden structure, Ganlan is highly flexible to suit different landforms. Therefore, it is widely used in the rainy and moist southern China. Apart from Tujia nationality, Ganlan wooden structure is used in traditional settlements of ethnic groups including Miao, Dong, Zhuang, Yao, Buyi, Dai, and Li. The Ganlan platform is where a whole family would live on. Typically, it includes a main room on the ground, "*huotangjian*" (room with fire-pit) as kitchen, and wing-houses with one side on the ground and the other side holding in the air. The family's daily life, dining and sleeping all take place on the Ganlan platform. Space under the platform is also used. In a mountain stilted building, the underfloor space is used to breed livestock and poultry, which is why "people can hear lambs bleat under stilted building" (*A Night at Mallard-nest Village*, written by Shen Congwen). In a waterfront stilted building, the platform extends to the middle of river, with columns standing on the riverbank. Space between columns can be small ports for boats or passageways for people. Stilted buildings written by Shen Congwen are mostly those built along rivers.

Tujia stilted buildings, as Ganlan architectures, can be divided into four types according to the construction size.

The first type is the most basic type. A row of three to five houses in a straight-line shape is built on a flatland. The main house is also referred to as "*zuoziwu*". The floorplan is either "one bright, two dark, three rooms" or "one bright, four dark, five rooms". In fact, *zuoziwu* is rarely lifted from the ground, unless the site cannot be flat. In this case, hanging feet are used to support *zuoziwu*, called "*diaoyitou*" (lift one side).

The second type is extending one lateral house that is called "*kanzi*" (den) in Hunan and western Hubei perpendicular to the main house on left or right side. This type is called "*yaoshitou*" (top of the key) as the main house and *Kanzi* are vertical to each other, looking like an old-style key. Due to the sloping terrain, *yaoshitou* is usually aerial or semi-aerial, functioning as an actual hanging foot.

The third type is to connect a *kanzi* on each side of the main house (i.e. to connect two *yaoshitou*) to create a U-shaped structure. Local people give this type two vivid names—"*cuojikou*" (open winnow) and "*yidantiao*" (shoulder pole). In essence, the third type can be seen as "*sanheyuan*" courtyard house (with three buildings).

The fourth type, enclosed "*tianjingyuan*" (patio courtyard), is formed by adding a reversed lodge outside of *cuojikou*. The fourth type is a high-standard settlement. Wealthy families may even build multiple *tianjingyuan* to form an estate, like Li's Mansion in Dashuijing.

Different types of stilted buildings, distinguished by their Ganlan wooden structure, make up Tujia settlements with varying shapes. From one generation to another, Tujia people have been living in these settlements. Tujia settlements are now "living specimens" of the ethnic collective living culture. To sum up, the Tujia settlement culture is as splendid as unique. Its process of fitting in the nature and forming a gregarious lifestyle, not only is an authentic and complete legacy, but also preserves its special features.

(Author: Li Xiaofeng)

大水井李氏庄园 Li's Mansion in Dashuijing
图片来源：作者自摄

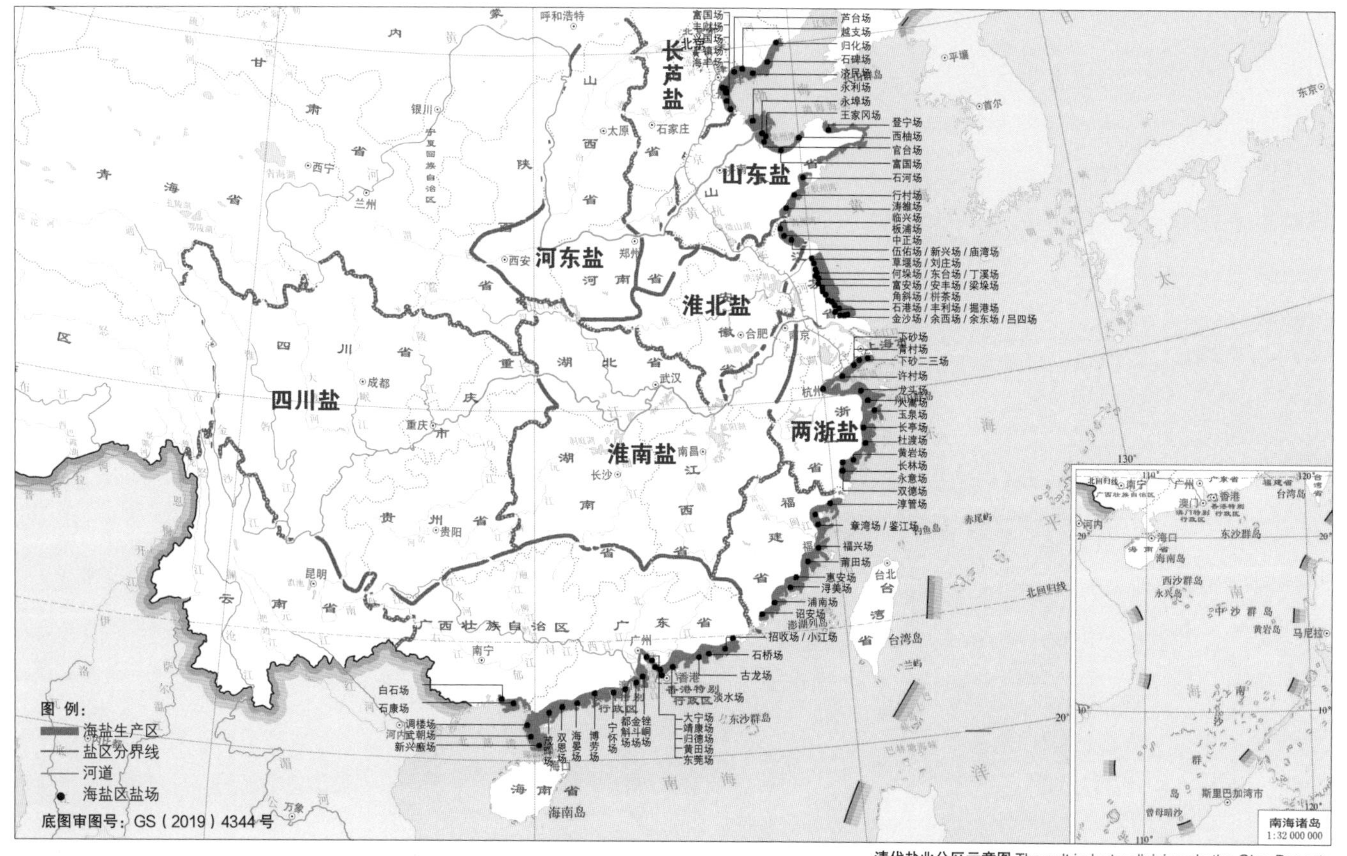

清代盐业分区示意图 The salt industry divisions in the Qing Dynasty
图片来源：赵逵．川盐古道：文化线路视野中的聚落与建筑 [M]. 东南大学出版社，2008.

1.4 川盐济楚与古盐道

人的生命离不开盐，人类活动要摆脱盐的束缚，必须依赖持续的长途运输。中国古代虽地域广阔，但食盐资源分布有限，也十分集中，除沿海地区外，内陆盐业资源仅集中分布于河东、四川、云南三处。其中川盐位于巴蜀地区的大山之中，周边湖南、湖北、四川、重庆以及贵州等地都不产盐，这些地区的人都需依靠川盐生存。因而古巴蜀人通过盐的贸易和周边地区的很多民族进行物质交换和文化、情感上的交流。盐为巴蜀这一原本非常封闭的区域打开了对外交流的通道。

近代，“川盐济楚”运动为巴蜀地区的经济、文化等带来了巨大的发展。抗日战争时期，国民政府被迫迁都重庆，失去了对东部富裕地区的税收控制。为解决战时资金短缺问题，弥补两湖地区的食盐空缺，国民政府发起了声势浩大的“川盐济楚”运动，以期利用川盐税收维持政府的正常运转。随着川盐大量运销长江沿岸，一批“富甲天下”的大盐商应运而生，川盐运输道路上更是千帆相竞、人流涌动，许多产盐古镇、运盐古镇应势而兴，迅速发展起来。现代众多巴蜀古镇多是在这一时期得到巨大发展，如纳水溪、庆阳坝、忠峒里、两河口彭家寨等。1949年后，共产党接管巴蜀，新的货币未到、旧的货币不可通行之时，“盐”由于不腐、易分割、易携带等特征，成为巴蜀地区的临时流通“货币”。

晶莹的“盐”在巴蜀地区是非常独特的存在，它不仅是持续运输能力的直接体现，更折射出了这一地区悠久的历史和神奇的文化。

（作者：赵逵）

鄂西宣恩县沙道沟镇彭家寨 Pengjia Village, Shadaogou in Xuan´en County in Western Hubei
图片来源：赵逵．川盐古道：文化线路视野中的聚落与建筑 [M]. 东南大学出版社，2008.

1.4 *Chuanyan Ji Chu* and the Ancient Salt Road

No one can live without salt. Therefore, continuous long-distance transportation is essential to free humans from salt's constraint. Although ancient China was vast in territory, salt distribution was both limited and unbalanced. Apart from coastal areas, inland salt resources were only available in Hedong, Sichuan and Yunnan. Among them, Sichuan salt is located in the mountains of Bashu area. None of the neighboring areas, including Hunan, Hubei, Sichuan, Chongqing and Guizhou, produce salt. All residents' survival in these areas relied on *Chuanyan* (salt from Sichuan). Via the trade of salt, ancient Bashu people exchanged material and communicated culture and emotions with people in neighboring areas. This originally isolated place was opened to the outside world by salt.

In recent history, Bashu area's significant development in economy and culture was brought by the "*Chuanyan Ji Chu*" (use salt from Sichuan to help Chu) movement. During the war of Resistance Against Japanese Aggression, the Nationalist government was forced to move its capital to Chongqing, thus losting revenue from wealthy eastern China. Hoping that tax from Sichuan salt could maintain its operation, the government initiated the robust "*Chuanyan Ji Chu*" movement to make up for wartime funding shortage and compensate for the Hunan and Hubei's salt scarcity. As a vast amount of Sichuan salt was transported for sale in the Yangtze Valley, a bunch of exceptionally wealthy salt traders emerged. The Sichuan salt transportation path was crowded with boats and people, which provided momentum for numerous old towns that either produced or were located along the salt transportation route to develop rapidly. Most of the ancient towns in Bashu, for example, Nashuixi, Qingyangba, Zhongdongli, Lianghekou, and Pengjia Village, prospered in this period. When the Communist Party of China took control over Bashu in 1949, salt was used as temporary currency when new currency hadn't arrived and old currency no longer circulated as it is non-decay and easy to separate and carry.

Glittering salt is a very special substance for Bashu. It is a straightforward demonstration of the area's transportation capabilities, as well as reflects the long history and spectacular culture in this area.

(Author: Zhao Kui)

两河口村模型
Models of Lianghekou Village

第二章 现场调研
CHAPTER 2 FIELD INVESTIGATION

两河口老街炭笔画 Charcoal drawing of Lianghekou old street
图片来源：张彤绘制

2.1 两河口村（中国 | 湖北）
Lianghekou Village (Hubei I China)

两河口村，位于湖北省恩施土家族苗族自治州宣恩县，是明清时期“川盐济楚”古盐道沿线的商贸村镇。百米古街保存完好，反映了典型的鄂西土家族苗族吊脚楼建筑的建构特征和商贸市村的聚落形态。由于人口流失，商贸凋敝，曾经兴盛的古村陷于衰败。

两河口村所在的沙道沟镇目前正在迎来发展的契机。以国保遗产彭家寨为代表的土家族村寨正在吸引世界的目光。2019年高速公路的通达为这条沉寂的盐道山谷带来大量客流。如何在迅猛的旅游业发展中，避免浅表性的消费，保存和发展国家级文化遗产的价值内核，使得沉睡的古村得到再生，这不仅是本次联合教学的主题，也是包括威尼斯在内的人类文化遗产共同面临的挑战。

Lianghekou Village, located in Xuan'en County, Enshi Tujia and Miao Autonomous Prefecture, Hubei Province, was a commercial village along the ancient Salt Road of "*Chuanyan Ji Chu*" during the Ming and Qing Dynasties. The one-hundred-meter old street, which typically embodies the tectonic characteristics of Tujia and Miao stilted buildings and the settlement pattern of commercial villages, has been well preserved. As a result of population loss and commercial decline, the once thriving ancient village is suffering from decay.

At present Lianghekou Village is embracing the opportunity of development. Tujia villages represented by the national heritage of Pengjia Village are attracting the attention from the world. The opening of the highway in 2019 is bringing numerous visitors to this hidden Salt Road valley. How to avoid superficial consumption and preserve the core value of the national cultural heritage in the rapid development of tourism, so as to regenerate the declined ancient village, is not only the topic of this joint teaching studio, but also the common challenge faced by human cultural heritages worldwide, including the city of Venice.

宣恩・两河口村老街鸟瞰
Overview of the Lianghekou Village in Xuan'en County

2.2 教学概述
Teaching Briefs

2019年4月至6月，在恩施土家族苗族自治州与宣恩县的支持下，受东南大学教育基金会与雅伦格文化艺术基金会资助，“两河口：土家盐道古村的再生”国际四校联合研学营在沙道沟镇彭家寨展开设计与研究。由东南大学、华中科技大学、威尼斯建筑大学、重庆大学等高校组成近50人的国际师生团队，共同探讨了衰败凋敝的盐道古村如何在即将到来的旅游业发展中获得再生契机，同时避免即时性和浅表性的消费对当地仍具生命力的文化遗产可能产生的磨蚀。

With the support of Enshi Tujia and Miao Autonomous Prefecture and Xuan'en County, funded by Southeast University Education Foundation and Fondazione EMGdotART, the four universities international joint studio, "Lianghekou: Regeneration of an Ancient Tujia Village along the Salt Road" was conducted in Pengjia Village, Shadaogou Town from April to June, 2019. A team of nearly 50 students and tutors from Southeast University, Huazhong University of Science and Technology, Università Iuav di Venezia and Chongqing University were gathered to explore how the declining ancient villages along the Salt Road can be regenerated in the upcoming tourism development, while avoiding the erosion of the vibrant local cultural heritage caused by instant and superficial consumerism.

教学海报 Poster of teaching

2.3 教学目标
Teaching Objectives

1. 证明与展现面对濒危脆弱的文化遗产，建筑学自身具有的力量和承担的责任。

2. 寻求一种建筑学的学科韧性：既不错失旅游经济带来的发展契机，同时保存和延续历史遗产的原真价值与文化认同。

1. To demonstrate the power and responsibility of architecture in the face of the increasingly endangered and fragile cultural heritage.

2. To seek the resilience of the architecture discipline: not missing the opportunities brought by tourism development, but also preserving and continuing the authenticity andcultural identity of historical heritage.

2.4 现场研学

On-site Teaching

42名学生被平均分为6组，每组7名学生来自不同学校。

6个小组依照自己的任务，在一周时间内进行针对性调研与分析。每组的调研工作分为2个部分：一方面着重进行老街的基础性测绘，从空间结构、比例尺度、材料技术等多角度进行全面而详细的信息收集与整理；另一方面每个小组须通过文献查阅、实地走访等方式在以下6个主题中选择一个进行专题研究调查。

42 students were divided into 6 groups. 7 students in each group are from different universities.

Following their tasks, the 6 groups had one-week time to conduct focused investigation and analysis. Tasks of each group are consisted of two parts. The first part targeted on the basic surveying and mapping of the ancient street. Students were asked to carry out comprehensive and detailed information collection and collation from different aspects, including the spatial structure, scale and size, material technology, etc.. The second part is to inquire into the chosen one of the following six themes for thematic research and investigation through literature reviews, field visits and other methods.

● 六个研究主题

Six research themes

1. 两河口村整体调研，包括新旧区域、地形和基础设施系统。
2. 从人类学、民族学、历史和经济学角度研究土家族。
3. 两河口村所在山区谷地的气候与地理条件、自然系统与资源。
4. 两河口村所在山区谷地的地域文化特征。
5. 土家吊脚楼及土家聚落的建构体系与建筑技术。
6. 恩施土家族苗族自治州和宣恩县的总体规划以及沿山谷区域的基础设施与社区服务系统。

1.Overall survey of Lianghekou Village including old and new areas, the topography and the infrastructure system.
2.Studying Tujia people from the views of anthropology, ethnology, history and economics.
3.The climate and geographic conditions, natural system and resources along the valley where Lianghekou Village is located.
4.Local cultural identities along the valley where Lianghekou Village is located.
5.Architectonic system and building techniques of Tujia stilted buildings and Tujia settlements.
6.General plan of Enshi Tujia and Miao Autonomous Prefecture and Xuan'en County as well as the systems of infrastructure and community service along the valley.

师生在彭家寨调研合影
A group photo of teachers and students of field investigation in Pengjia village

师生在两河口老街现场调研 1
Teachers and students are conducting on-site research on Lianghekou old street 1

学生在彭家寨现场听老师讲解
Students are listening to the teacher's explanation on site in Pengjia Village

与当地村民交流访谈
Communicating and interviewing with local villagers

师生在两河口老街现场调研 2
Teachers and students are conducting on-site research on Lianghekou old street 2

各组同学以手绘、速写、图解、照片、视频、笔记等多种方式对测绘成果与专题研究进行综合展示，并讨论了初步的设计思考，以应对老街未来发展的可能性。

Each group made a comprehensive display of the various surveying and mapping achievements and research topics in multiple ways, such as drawings, sketches, diagrams, photography, video, written notes, etc.. Some of them proposed preliminary design concepts to cope with the possibility of future development of the historical streets.

师生在两河口老街现场调研 3
Teachers and students are conducting on-site research on Lianghekou old street 3

师生在彭家寨现场讲解
Teachers and students are expounding on site in Pengjia Village

师生在宣恩县档案馆调研
Teachers and students are conducting research at the Xuan'en County Archives

中意师生在两河口小学汇报评图
Chinese and Italian teachers and students are making presentation and review in Lianghekou Primary School

国际研学营期间，中意四校师生和当地村民共同体验土家乡土生活。两河口村的生活和学习的经历给我们留下了深刻的印象：与当地学校小学生打成一片，听老街中留守的老人唱山歌，和来自不同文化背景的互相交流。

During the international joint studio, Chinese and Italian teachers and students from the four universities experienced the Tujia rural life with local villagers. We were deeply impressed by the living and learning experience in Lianghekou Village: having fun with the local primary school students, listening to the elders in Pengjia Village singing folk songs, and communicating with each other from different cultural backgrounds.

中意师生在两河口小学合影
A group photo of Chinese and Italian teachers and students at Lianghekou Primary School

● 场地测绘的模型与图纸
Models and drawings of the site surveying

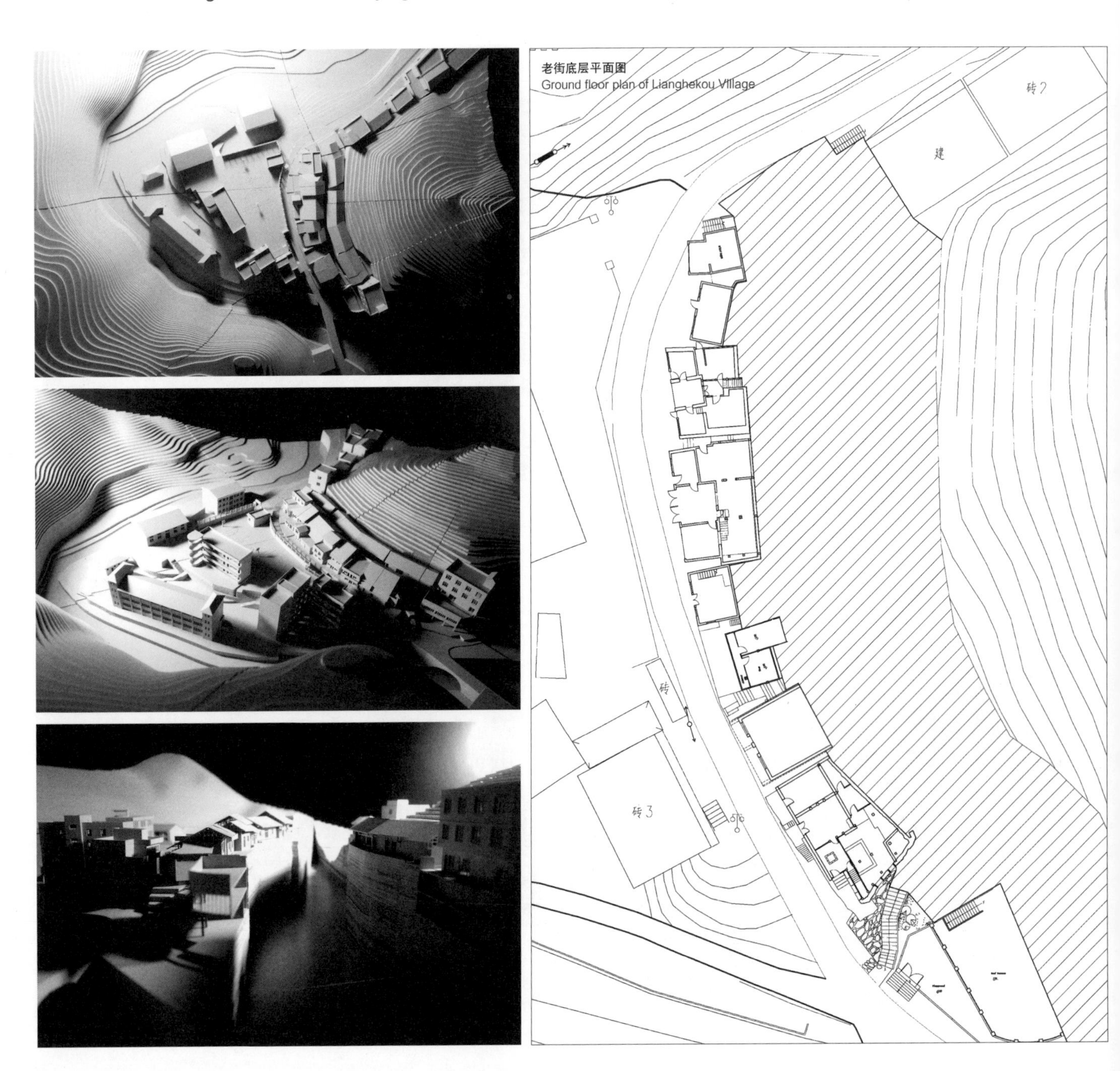

老街一层平面图
First floor plan of Lianghekou Village
老街二层平面图
Second floor plan of Lianghekou Village
建
砖2
砖
木
砖3
学
水泥
10 m

龙潭河与酉水河河口山谷地形分析

Valley topographical anaylsis of Longtan and Youshui Rivers Estuary

图片来源：威尼斯建筑大学小组学生绘制

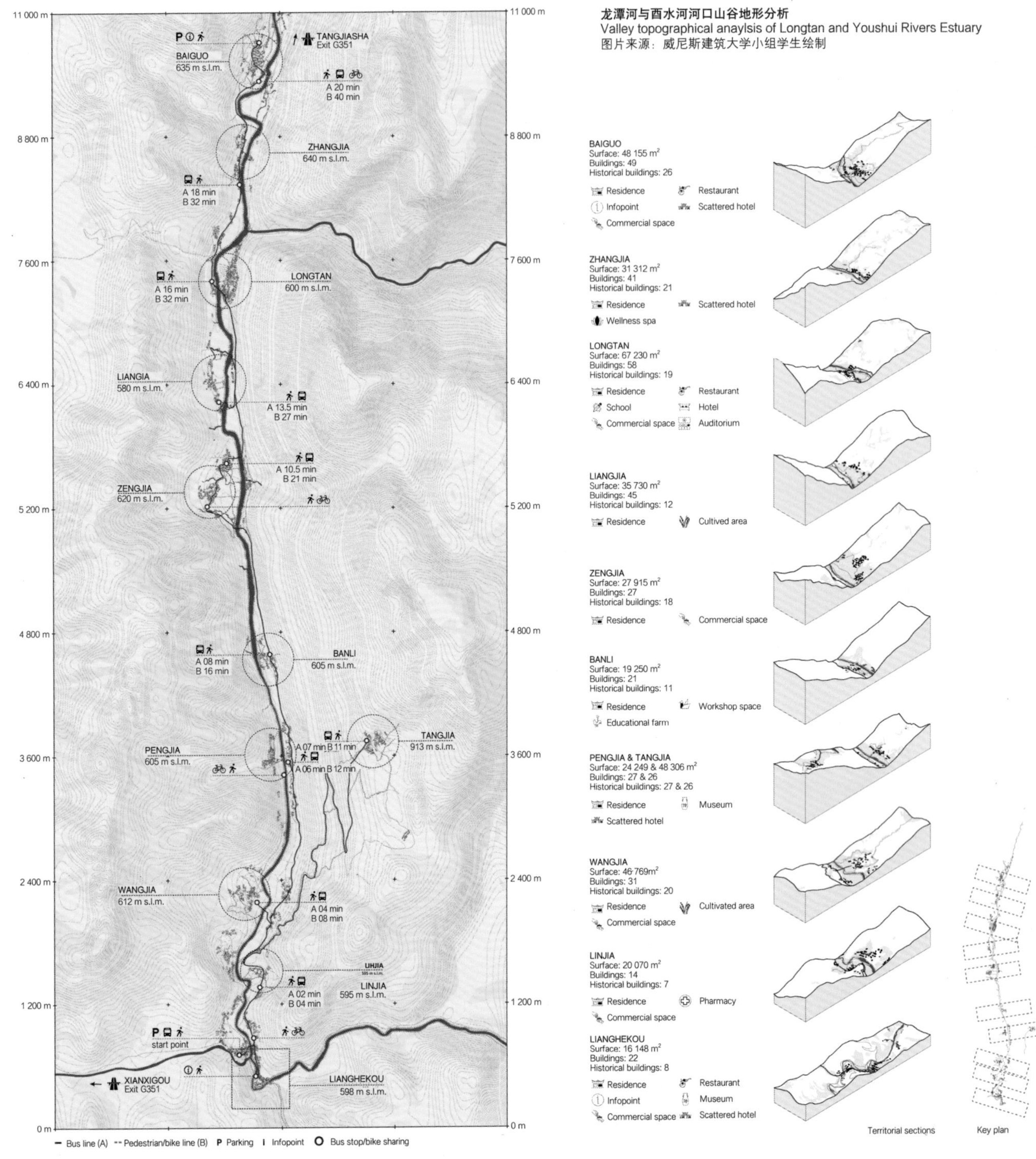

两河口河谷植被与场地视线分析
Vegetational and site viewing anaylsis of the Lianghekou valley
图片来源：彭欣怡、王沛赉等绘制

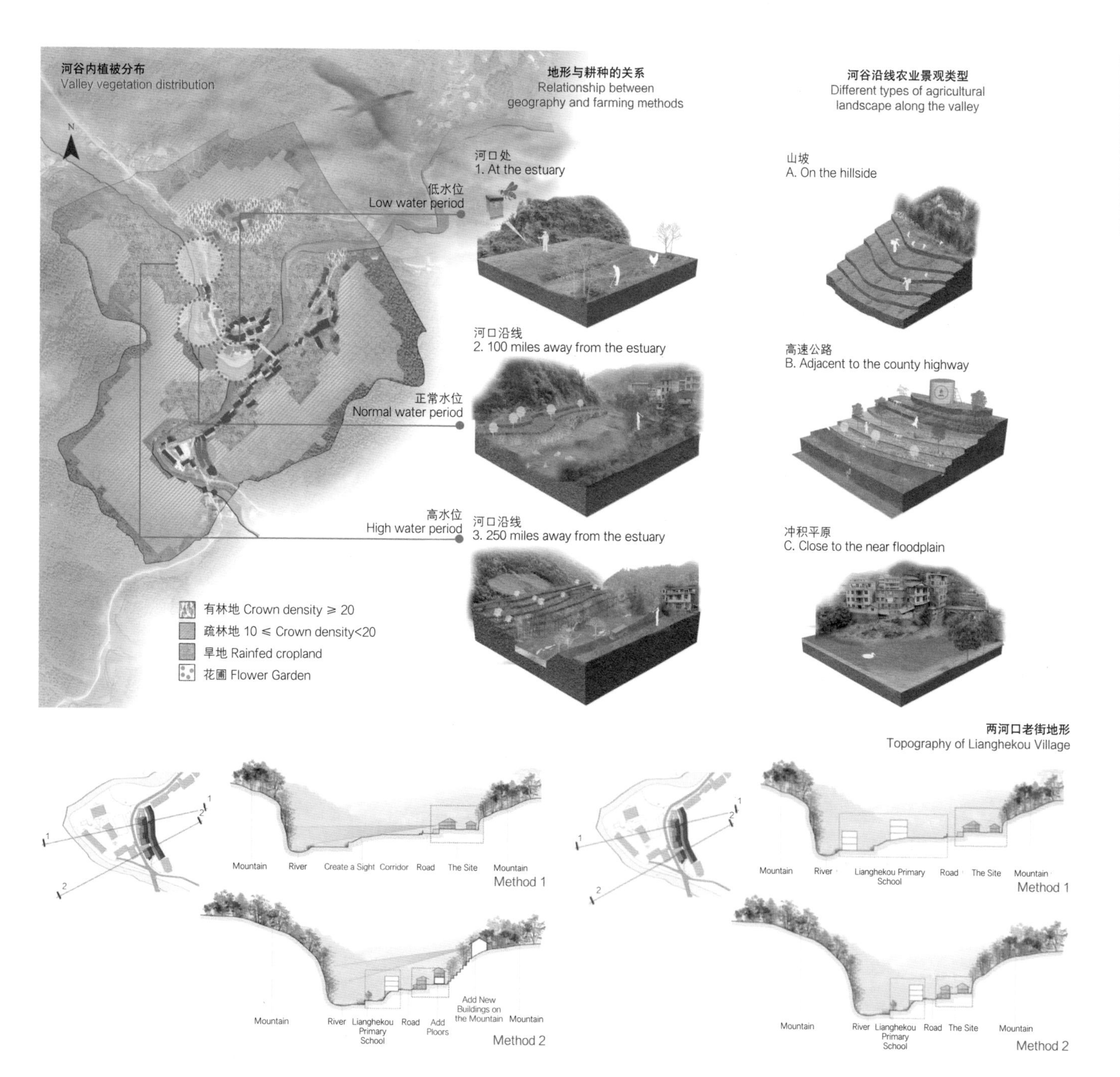

两河口老街地形
Topography of Lianghekou Village

手绘考现记录

Hand–drawing records

宣恩彭家寨土家吊脚楼速写 Sketch drawing of Tujia stilted building, Pengjia Village, Xuan'en

图片来源：徐涵绘制

土家吊脚楼穿斗构架 *Chuan–dou* type frame of Tujia stilted building

图片来源：徐涵绘制

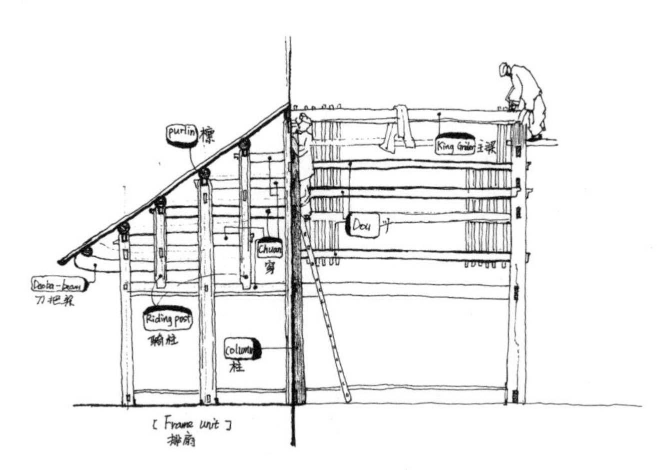

两河口老街速写 Sketch drawing of Lianghekou old street

图片来源：马雨萌绘制

● 空间尺度分析
Spatial scale analysis

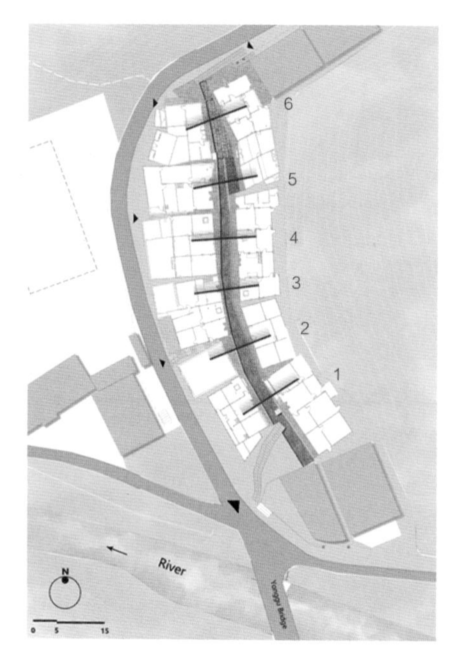

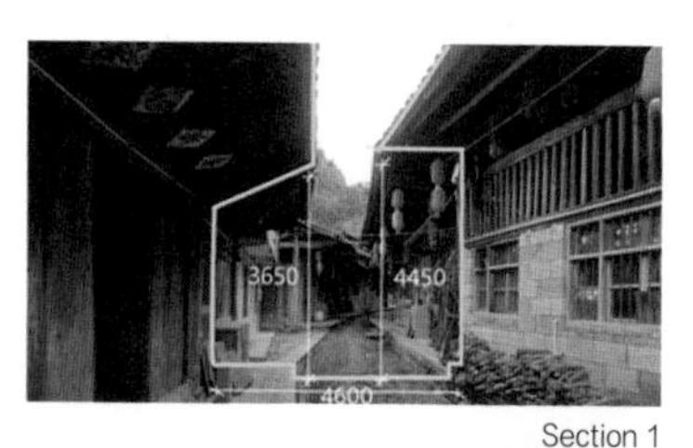

Section 1

Section 4

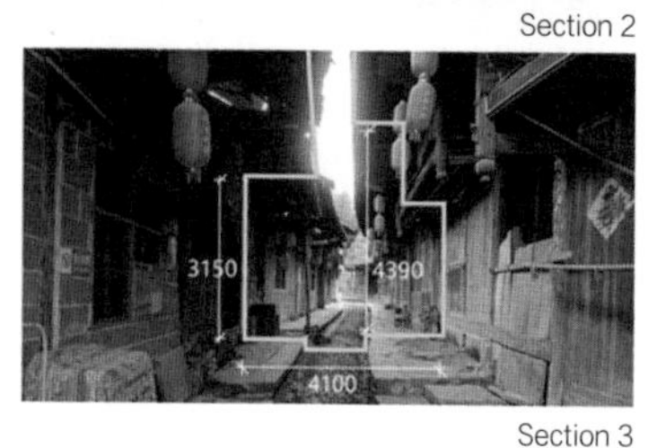

Section 2

Section 5

Section 3

Section 6

老街的空间序列尺度 The spatial sequential scale of the old street
图片来源：马雨萌、刘文玉等绘制

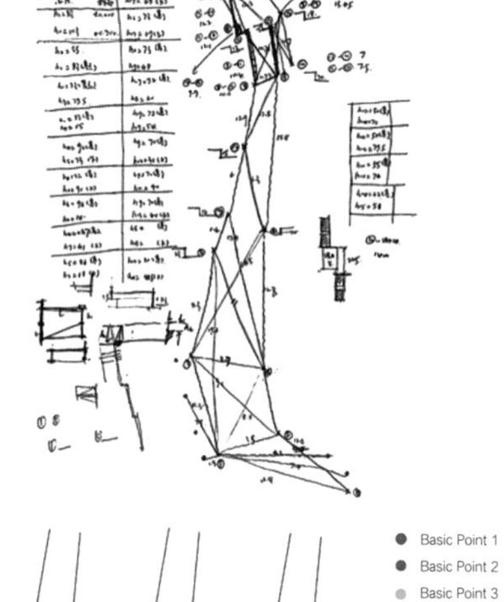

老街的测绘记录 Surveying records of the old street
图片来源：Anna、Elisa 等绘制

old rubble with cement cover
large pebble
new slabstone
cement with moss
new slabstone
old slabstone
old stone with cement cover and moss
old stone
hole

Basic Point 1
Basic Point 2
Basic Point 3

Step 1
Step 2
Step 3

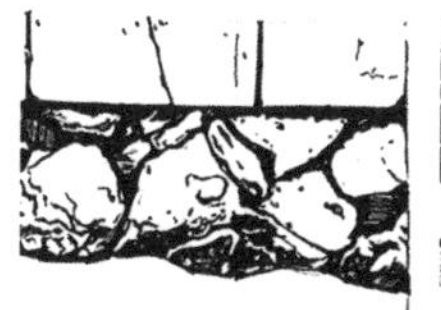

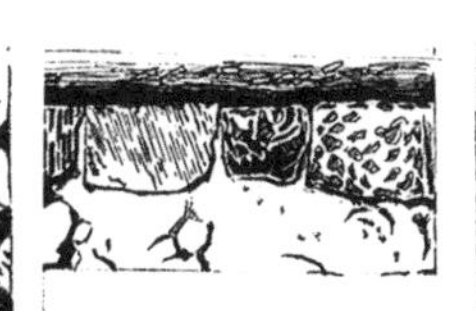

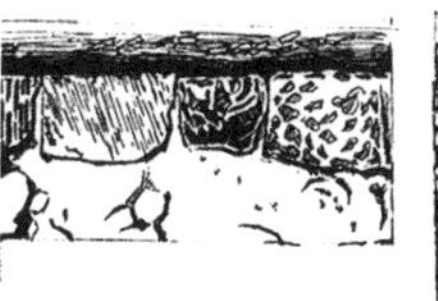

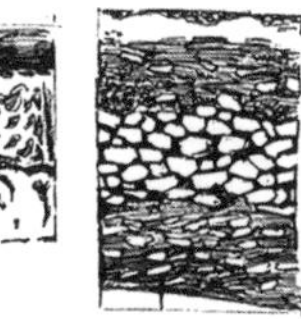

● 土家民居的功能变迁
Evolution of the function in Tujia residential buildings

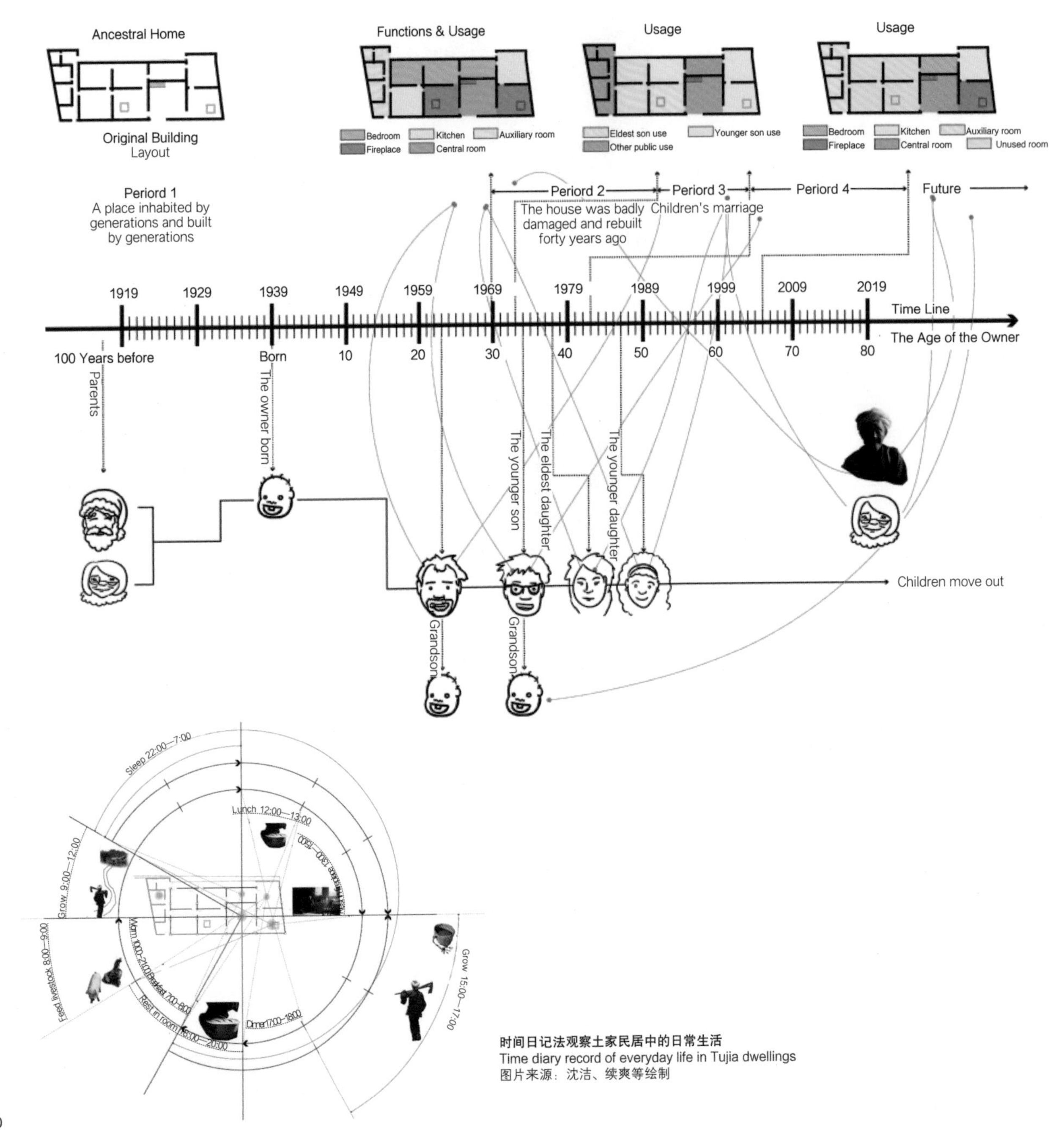

时间日记法观察土家民居中的日常生活
Time diary record of everyday life in Tujia dwellings
图片来源：沈洁、续爽等绘制

● 老街生活场景再现

Representing living lifestyle of the old street

在为期一周的联合教学过程中，采用现场田野调查的调研方法，深入体察老街土家族吊脚楼建筑的建构特征和村民在吊脚楼空间中的生活习惯，以图像的方式生动记录，作为后续设计活化的参考基础。

During the one-week joint teaching studio, an on-site investigation was adopted to record the construction characteristics of the Tujia nationality's stilted buildings in the old street and the living habits of villagers in the space. Images were used to vividly record and serve as a reference basis for coming architectural design.

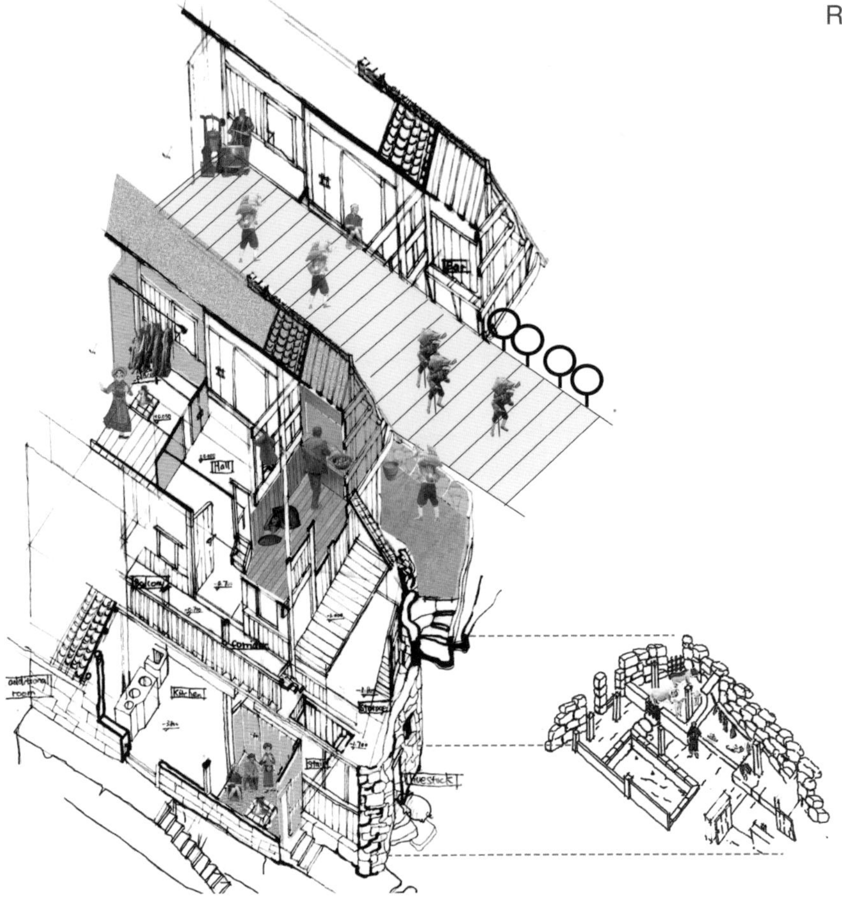

两河口老街土家族民居生活场景记录

Drawings of the living scenes of Tujia residential buildings in Lianghekou old street

图片来源：徐涵、李晨等绘制

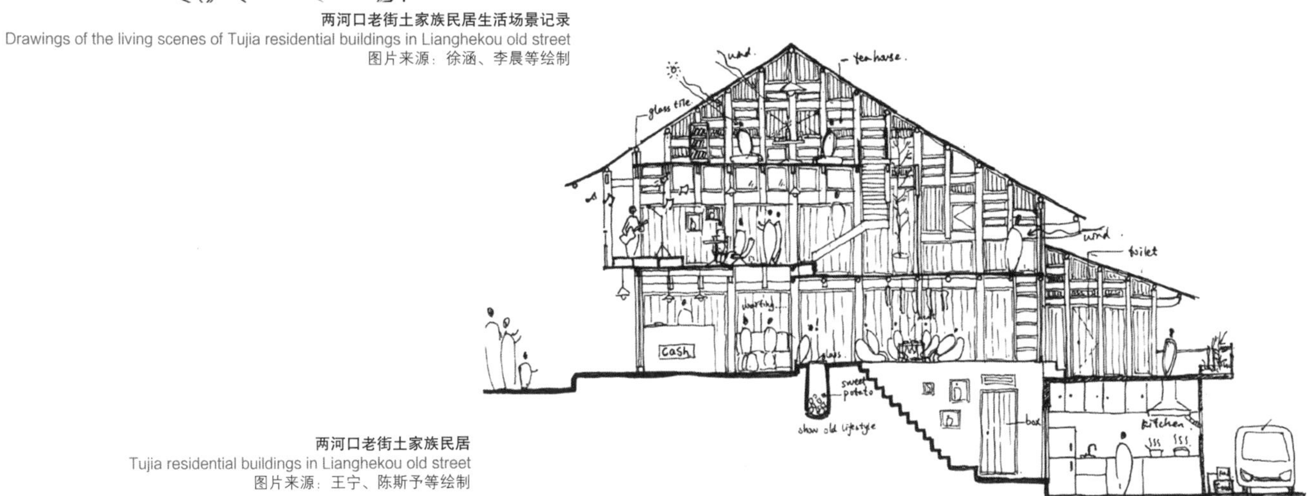

两河口老街土家族民居

Tujia residential buildings in Lianghekou old street

图片来源：王宁、陈斯予等绘制

● 两河口村土家吊脚楼建构体系研究

Architectonic system of Tujia stilted buildings in Lianghekou Village

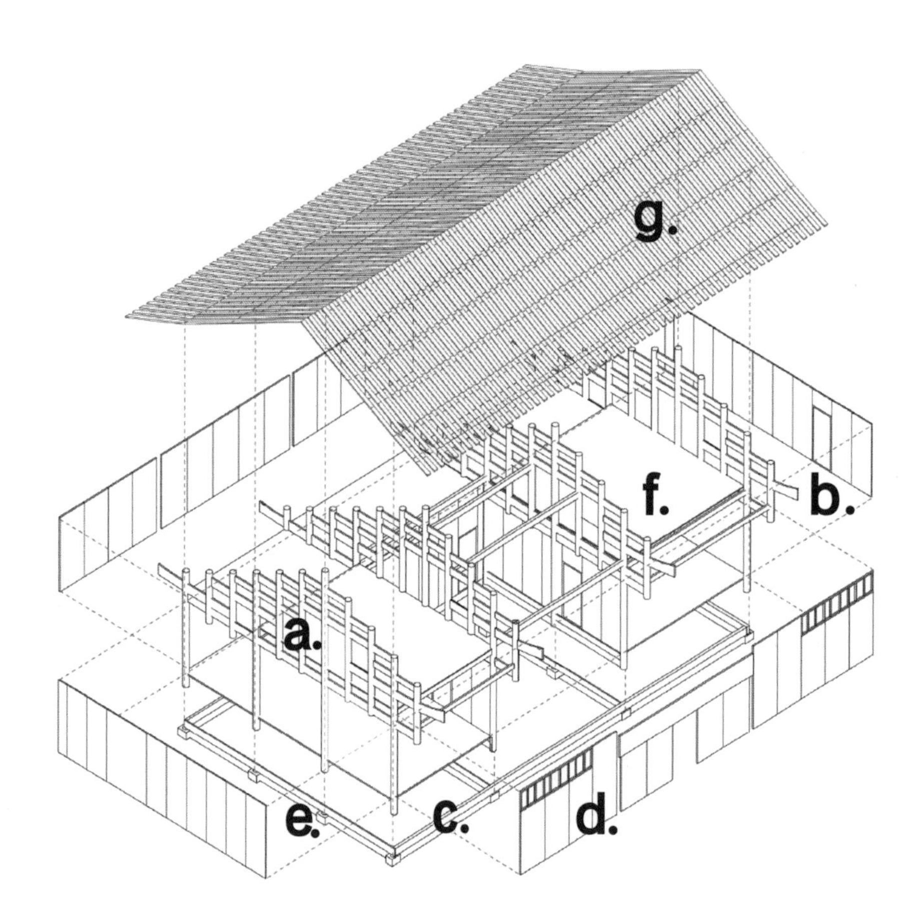

a. Frame 构架

b. Spatial distribution 空间布局

c. Basement 基础

d. Walls and openings 墙体门窗

e. Materials 建造材料

f. Floors 楼板

g. Roof 屋面覆盖

a. Frame

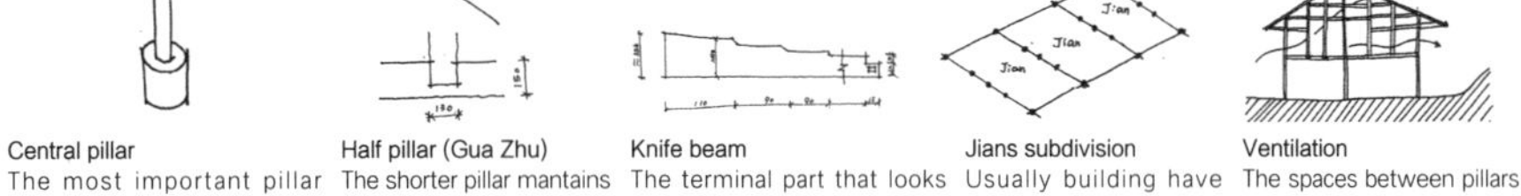

Central pillar
The most important pillar toghether with the corner one; a red paper is usually sticked to it with luck words.

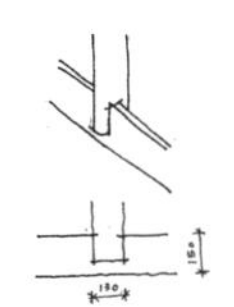

Half pillar (Gua Zhu)
The shorter pillar mantains the space integrity.

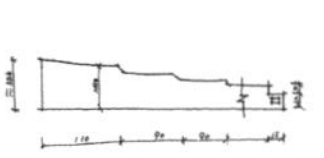

Knife beam
The terminal part that looks like a knife: this beam passes throught two or more pillars and hooks with them.

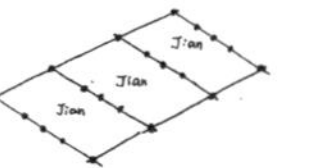

Jians subdivision
Usually building have three jians, the central one is the ritual part.

Ventilation
The spaces between pillars and beams in the frame leaves the wind to across the structure.

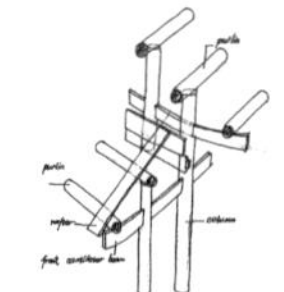

Side eave to expand space

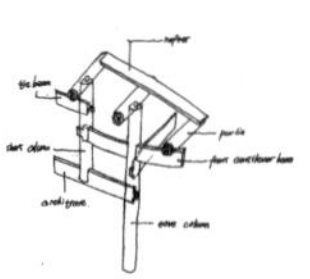

Joint pillar-balcony

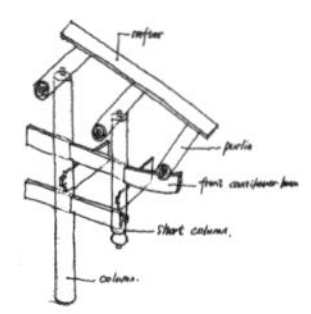

Double front beam

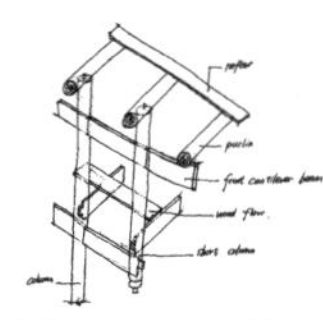

Balcony spaces craetion

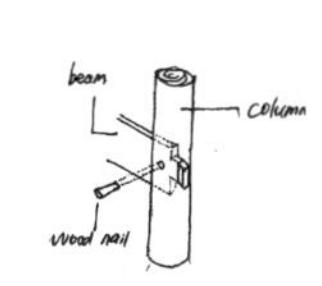

Basic joint

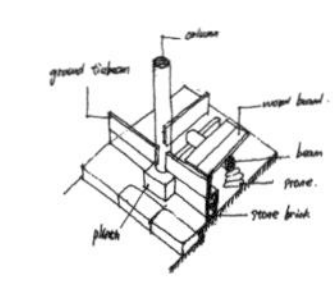

Joint pillar-ground

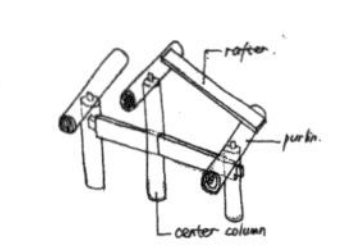

Central beam joint

b. Statial distribution

Private space
This area is usually elevated from the ground level by a wooden floor. It defines the owner's private spaces.

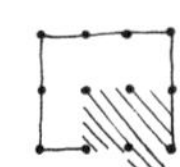

Commercial space
This area is usually defined by an inside part on the footpath level; the exterior part could be characterized by a wooden shelf.

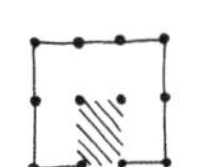

Sanctuary area
This is the place where owners pray; in ancient times they prayed in the direction of the central pillar.

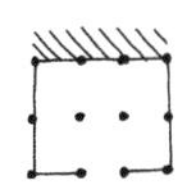

Backyard
Area behind the building is used for storage. It can also be used like places for animals and vegetables garden.

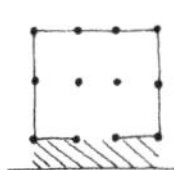

Walkable space
It's an exterior space defined between two limits, usually protected by a rooftop.

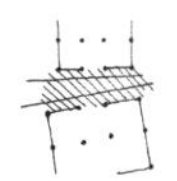

Commercial road
In ancient times all the commercial spaces look out on this transports way.

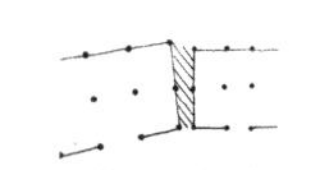

Spaces between buildings
In Tuija tradition there are not common walls between two different properties.

c. Basement

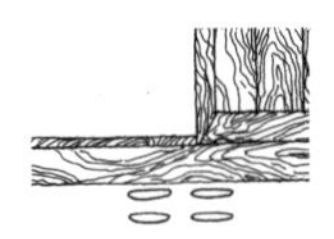

Concrete basement

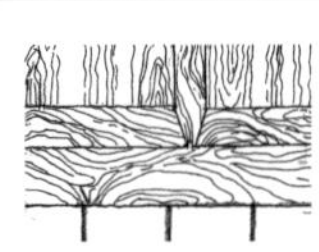

Concrete blocks basement

Stone basement

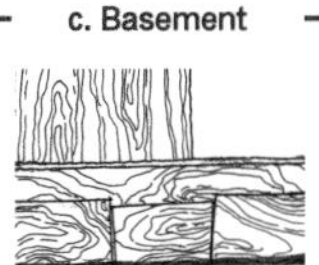

Wooden basement

Irregular stone plinth

Regular stone plinth with stone basement

Regular stone plinth with slab stone basement

d. Walls and openings

Property limit
The walls divide the property of the building and protect property from bad wheather.

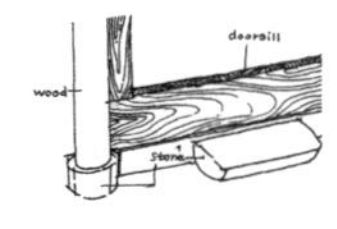

Entrance limit
Doorstep implies a sign of modesty from those who enter in the house from the outside.

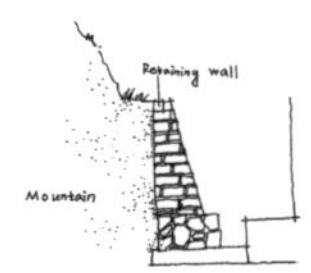

Retaining wall
It is used to create a planar spase in a sloping topography. It also collects the water that comes from the mountain.

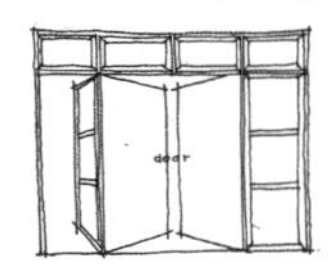

Openable wall
We can find it in a modern building.Private and public spaces are more connected.

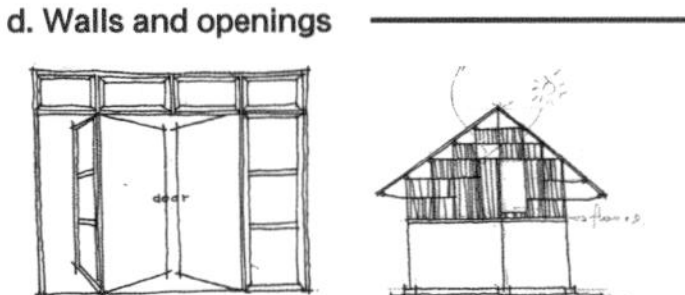

Holes for upper entrance
Openings on the walls of the second floor allow access to the upper part of the building.

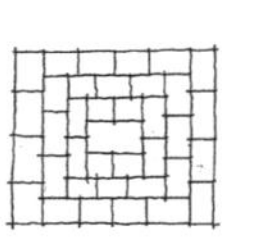

Window
Holes with the functions of ventilation and lighting.

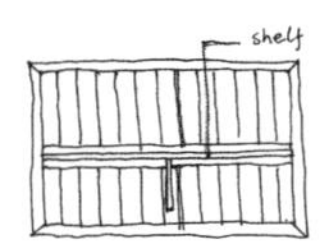

Commercial window
It's a big window with a shelf that serves as a bar counter.

两河口村土家民居建构体系谱系表 Genealogy of the architectonic system of Tujia residential buildings in Lianghekou Village
图片来源：Davide、Carton 等绘制

d. 1 Founctions

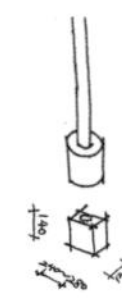

Pillar plint
The structural pillars are placed over plinths without any joints.

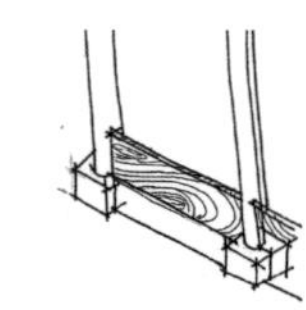

Wall basement
The under-wall structure, usually made of stone, raises the building from the ground for water and humidity problems.

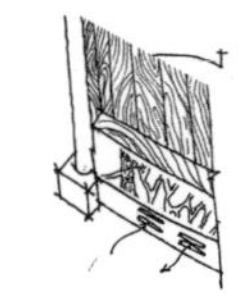

Ventilation
Sometimes the basements present some holes for ventilation and water drain. They usually are images of local money icon.

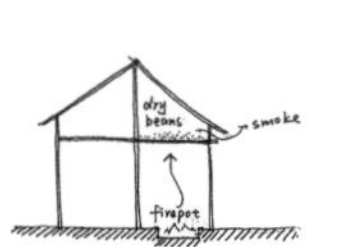

Fire pot
During the years, we can find the fire pot in different places in the traditional houses. This is caused by the building enlargement.

Full basement
In some ancient villages, like Lianghekou, the basement response to monumental volume.

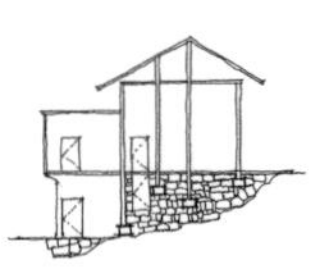

Livable basement

Hidden or absent plinth

e. Materials

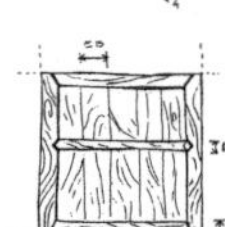

Wooden wall
The principal structure is filled up of wooden boards.

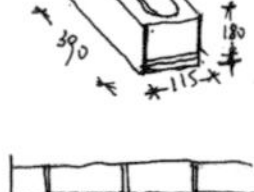

Concrete block wall
Modern materials used for addiction in recent times.

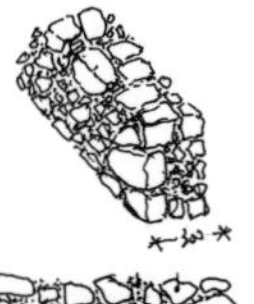

Mixed wall
Composed by clay, stone and rice hull.

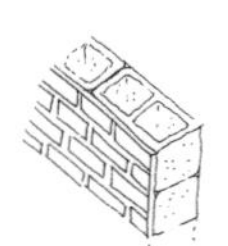

Tiled wall
Covering kind diffused in all the Hubei area.

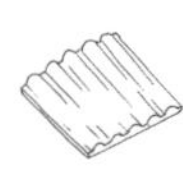

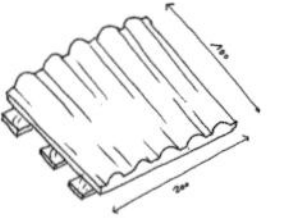

Plastic tiles

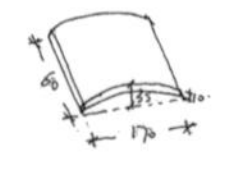

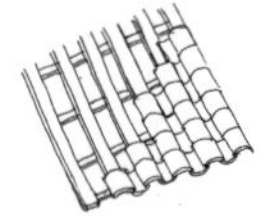

Traditional tiles

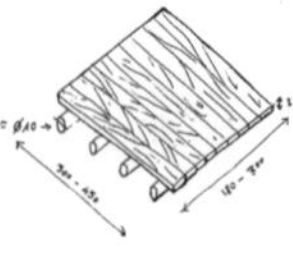

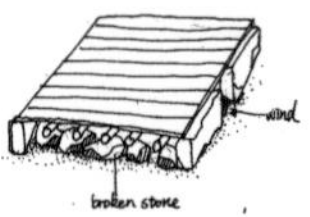

Planar ground floor
The private space is defined by the wooden area.

f. Floors

Storage
In old times the second floor was used to store the rice and let it dry.

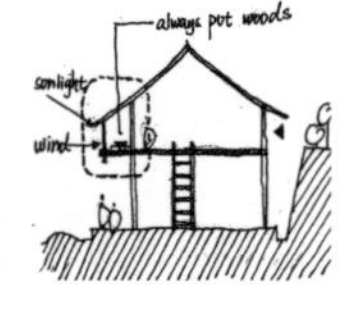

Balcony
The extension of the second floor created balconies.

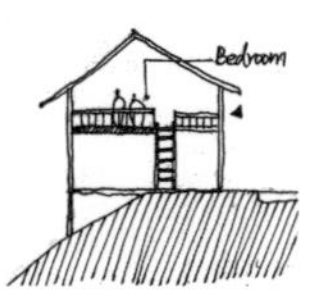

Private room
During the years the storage floors were closed up to create livable space.

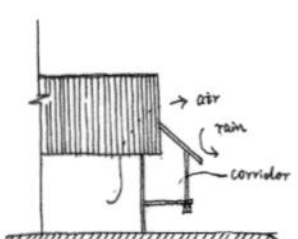

External corridor
The addiction of shelter gave the possibility to use space at different levels.

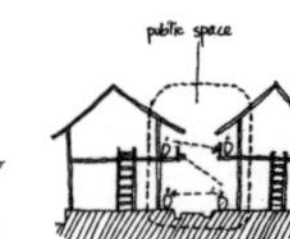

Private/public space
The relation between houses creates relation between people.

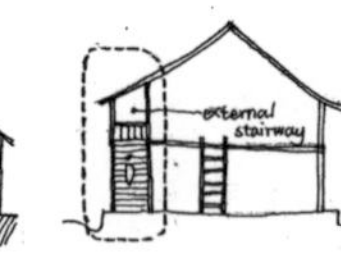

Stairs
Existence of many stairs, like interior/exterior with different inclinations.

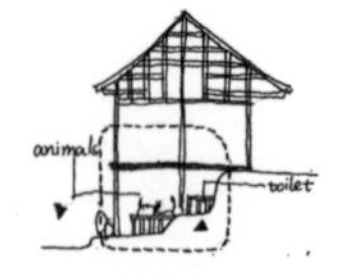

Underground level
The underground level in old times was used to accomodate animals.

g. Roof

Rain drainage
angle of roof defined by tradition, rainfall and building orientation.

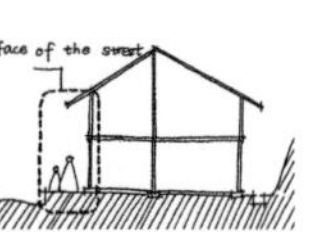

Covered passages
The extension of the roof defines social spaces sheltered from the rain.

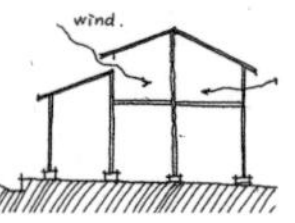

Ventilation
Roof discard is determinated by the need of second floor ventilation, usually dedicated to the rice storage.

Sun light
Some holes on the roof were designed to let the sunlight to light up many spaces of the house.

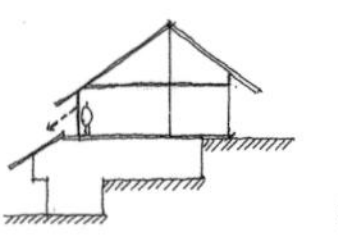

View point
Roof discard creates some observation decks.

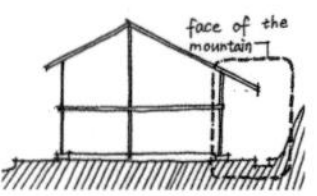

Private spaces
The extension of the roof on the back of the house creates different not-collective space.

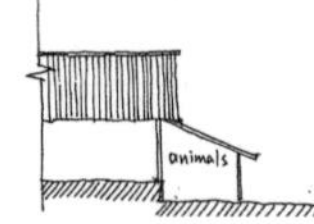

Home extensions
New roofs were originally created by the necessity of new additions, caused by the growth of the family core.

第三章 规划与设计
CHAPTER 3 PLANNING AND DESIGN

Group 1/SEU 第一组 东南大学

3.1 木构变色龙：从日常生活到节日庆典

Wooden Chameleon: From Daily Life to Festival

学生：隋明明、陈斯予、闫宏燕、李心然
Students: Sui Mingming, Chen Siyu, Yan Hongyan, Li Xinran

指导老师：张彤、王川
Tutors: Zhang Tong, Wang Chuan

中国南方传统的干栏木构建筑的承重体系与围护体系分离，空间和结构呈现出灵活性、可变性和适应性。与中国各地乡土聚落一样，土家族村寨的街道是集体生活的重要场所。干栏木构的可变性与灵活性，使得公共空间展现出迷人的适应性，其形态随着人们的生活方式而改变。现场调研观察到土家族生活的三种典型状态：日常、赶集与庆典。它们需要的公共空间是不同的。该设计将两河口村作为一个整体建构体系，研究其固定与可变的不同组成构成，利用其可拆装的构造方式，以一个结构体系的拓朴变形，适应与承载日常生活、乡村集市和节日庆典的不同空间需求。

The traditional Ganlan wooden architectures in South China, with the separation of load-bearing systems and envelope systems, present flexibility, variability and adaptability in space and structure. Like vernacular architecture all over China, the streets of Tujia villages are important places for collective life. The changeability and flexibility of Ganlan structure make the public space show a charming adaptability, whose form changes with the change of people's lifestyle. Three typical states of Tujia life were observed in the field investigation: daily, fair and celebration. They require different public spaces. This design studies Lianghekou Village as a whole construction system, with its fixed and variable different components. By using its dismountable construction mode, the topological deformation of a structural system adapts to and carries different spatial requirements of daily life, rural markets and festival celebrations.

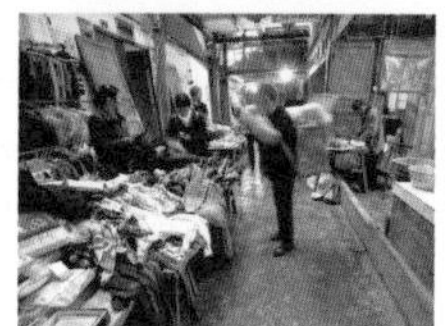

日常生活 Daily life

乡村市集 Rural markets

节日庆典 Festival celebrations

Features of Chinese traditional wooden structure
中国传统木构特征

- Flexibility 灵活性
- Variability 可变性
- Adaptability 适应性

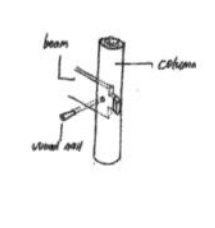
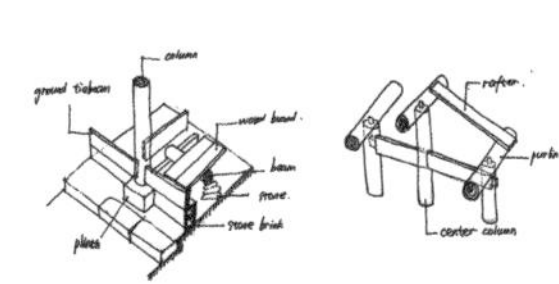

基本连接构造 Construction of basic joint

木构建造体系的分析
Analysis of wooden construction system

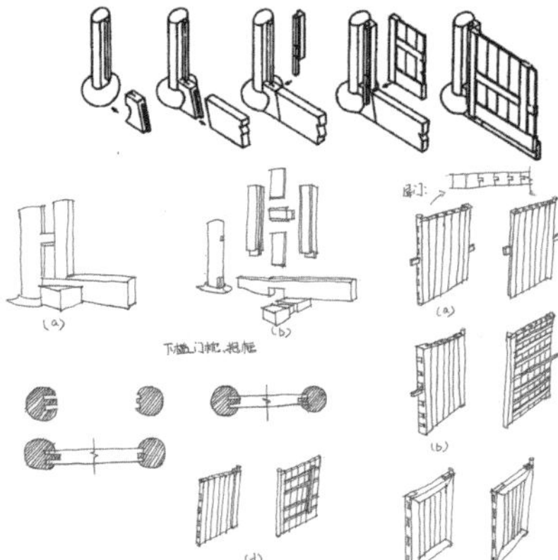

细木工榫卯连接节点
Joinery joints with mortise and tenon joints

● 老街中建筑要素分析
Analysis of architectural elements in the old street

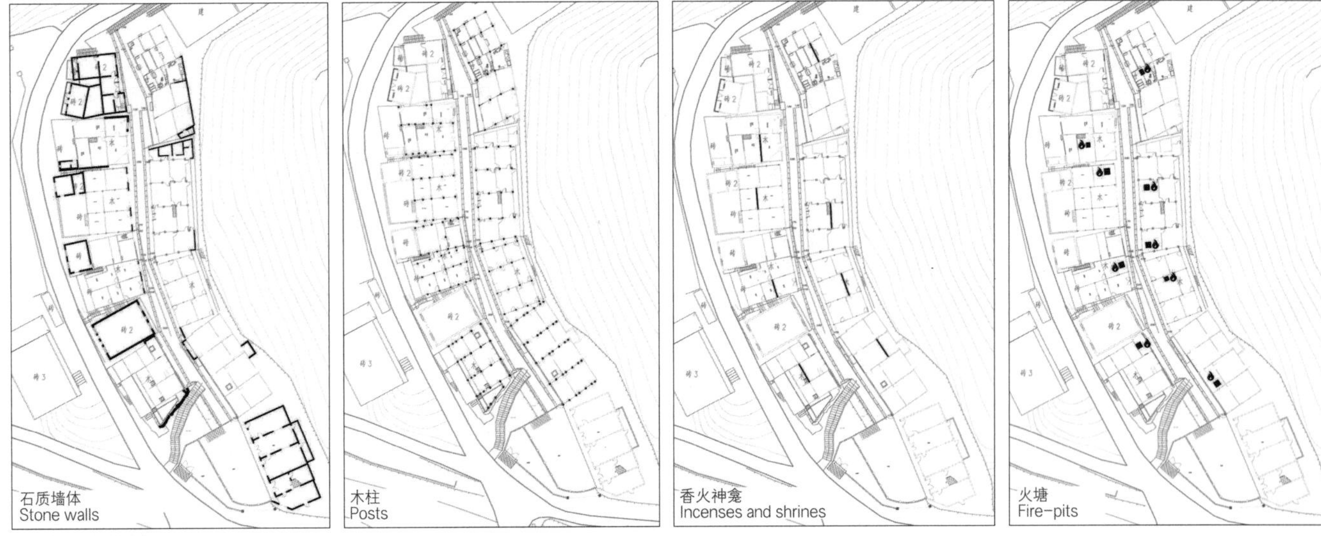

保留的不可移动要素
Retained immutable elements

调整的可变要素
Adjusted variable elements

● 老街中三种生活空间场景
Three living spatial modes in the old street

日常模式
Daily mode

日常模式下老街底层平面的功能分析
Functional analysis of the ground floor plan of the old street under the daily mode

集市模式
Market mode

集市模式下老街底层平面的功能分析
Functional analysis of the ground floor plan of the old street under the market mode

庆典模式
Festival mode

庆典模式下老街底层平面的功能分析
Functional analysis of the ground floor plan of the old street under the festival mode

● 日常模式
Daily mode

改造后的一层平面 The renovated first floor plan

老街最常见的面貌是村民日常生活状态下的使用空间。整条街道，从老街一端延伸至末尾的幼儿园，呈线性布局特征，形成一个由传统木构建筑单体组成的连续木构空间。这一空间情景下，需要进行乡村基础设施的完善与更新。设计补充了街道中游客卫生间、垃圾处理站、游憩凉亭等公共服务空间，并恢复了幼儿园立面风貌。

改造剖面 The renovated section

The most common appearance of old streets is the use of space in the daily life of villagers. The entire street, extending from one end of the old street to the kindergarten at the end, presents a linear layout feature, forming a continuous wooden space composed of traditional wooden building units. In this spatial scenario, rural infrastructure needs to be improved and updated. The design has supplemented the public service spaces in the street, such as tourist bathrooms, garbage disposal stations, recreational pavilions, and restored the facade of kindergartens.

● 集市模式
Market mode

改造后的一层平面 The renovated first floor plan

土家族聚落每月逢农历二、五、八日有集市赶场的习俗。因此，乡村集市成为土家族商贸聚落中的一个重要空间使用模式。土家集市不仅吸引了周边村落的赶集村民，也常常汇聚了大量途经此地的商队小贩，骡马成群。在这一特殊聚集性生活场景中，老街中原有民居的大门门板和木窗隔板被灵活拆除，形成售卖的橱窗和店面空间。民居首层变为临时性的商业空间，陈列着琳琅满目的商品。

改造剖面 The renovated section

Tujia settlements have the custom of going to the market on the second, fifth and eighth days of every lunar month. Therefore, the rural market is also an important space in the Tujia business settlements. The Tujia market not only attracts villagers from surrounding villages, but also often gathers a large number of caravans and vendors passing through the area, with mules and horses flocking together. In this special gathering life scene, the front door panels and wooden window partitions of traditional houses in the old street can be flexibly removed to form a display window and storefront space for sale. The first floor of the residential buildings has become a temporary commercial space, displaying a dazzling array of goods.

● 庆典模式
Festival mode

改造后的一层平面 The renovated first floor plan

老街棚布和排水系统设计
Canopy and drainage system design in the old street

改造剖面 The renovated section

在重要的祭祀节庆日，土家族人会举行摆手舞、长桌宴等民俗庆典活动。这一节日庆典模式需要更多的公共空间来容纳歌舞仪式活动。设计方案通过在老街上空增加一个五彩经幡形成遮蔽，并利用木构建造的灵活性，将建筑一层的木隔板尽数移除，使得空间成为一个极致状态下的敞厅。

On important sacrificial festivals, Tujia people will hold folk celebration activities such as hand-waving dances and long table banquets. This festival celebration mode requires more public space to accommodate singing and dancing ceremony activities. The design scheme involves adding a colorful prayer flag over the old street to form a shelter, and utilizing the flexibility of wooden construction to remove all the wooden partitions on the first floor of the buildings, making the space an ultimately open hall.

Group 2/SEU 第二组 东南大学

3.2 土家民艺村：公共空间与传统吊脚楼活化

Tujia Artware Village: Regeneration of Public Space and Traditional Stilted Buildings

学生：徐涵、马雨萌、王蓉
Students: Xu Han, Ma Yumeng, Wang Rong

指导老师：张彤、王川
Tutors: Zhang Tong, Wang Chuan

设计从两河口村与其所在山谷中九寨的区位关系出发，期望打造一个与周边其他村寨功能互补又独具特色的空间节点以吸引来访者。设计将挖掘、展示、传承、培训土家族丰富多彩的民族手工艺作为老街新的功能定位，在唤醒濒临失传的民艺，增强地区文化自信的同时，发展新的展陈、民宿、餐饮接待等服务性产业，以激活衰败的古村，引导外出村民重新回乡就业。我们选择改造两个现有建筑。一个建筑是大凉亭（土家族大亭子），它是一座敞开的木亭，矗立在村庄北部边缘的一所小学对面。改造的亭子旨在成为一个多功能的公共空间，其灵活的布局融入村庄拓扑变形的公共空间中，成为社区各种公共生活模式的焦点。该设计继承了土家族的木结构传统，在模块化结构体系和典型的木结构中采用了传统技术。另一个建筑是位于街道南端的土家织锦作坊——西兰卡普工坊。工坊最初的照明过于昏暗，无法满足新的功能要求，而吊脚楼结构造成了许多内部高度差异。为了解决这些问题，西兰卡普工坊增加了一个中庭和一个室外晾晒的通风平台，创造出具有土家特色风情的独特体验空间。

In view of the geographical relationship between Lianghekou Village and nine villages in the valley, the design aims to create a space with complementary and unique functions with other surrounding villages to attract visitors. The new function of excavating, displaying, inheriting, and training Tujia's rich and colorful folk handicrafts can arouse the awareness of traditional techniques and enhance regional cultural confidence. Combined with new services such as exhibitions, homestay and catering reception, it can regenerate the decaying ancient village and attract the original villagers to go back to their hometown for employment. We chose to remould two existing buildings. One is Tujia Grand Liangting (grand Tujia pavilion),which is an open wooden pavilion that stood opposite a primary school at the northern edge of the village. The remoulded pavilion was intended to be a multi-functional public space, with its flexible layout integrated into the topologically deformed public spaces of the village, and became the focus of various modes of public life in the community. This design inherited the Tujia tradition of timber construction by employing traditional techniques in the modular structural system and typical wooden structures. The other structure was the Silankarp (vernacular Tujia brocade) workshop located at the southern end of the street. The original lighting of the workshop was too dim for the new functional requirements, and the stilted structure created many internal height differences. To address these issues, an atrium and an outdoor brocade ventilation deck were added to the Silankarp workshop to create a unique experiential space.

土家民艺村方案场地模型 Site model of the Tujia Artware Village

土家民艺村一层平面 First floor of the Tujia Artware Village

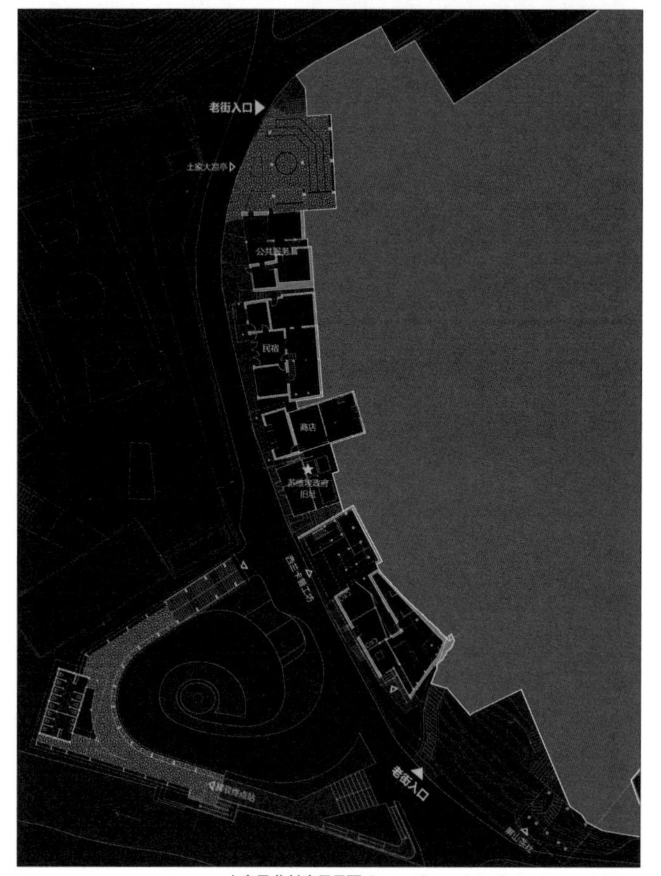

土家民艺村底层平面 Ground floor of the Tujia Artware Village

● 土家大凉亭
Tujia Grand Liangting

土家大凉亭透视 Perspective of Tujia Grand Liangting

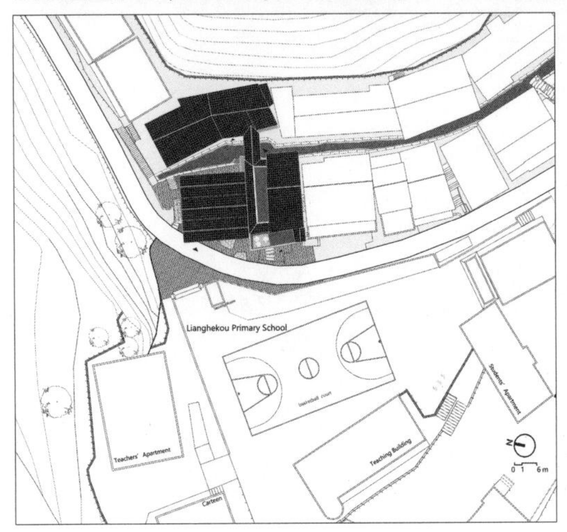

左：土家大凉亭模型 Left: Model of Tujia Grand Liangting
右：土家大凉亭总平面 Right: Master plan of Tujia Grand Liangting

老街复兴采用两个策略：
一是村庄整体公共空间重塑与品质提升，包括改造村口幼儿园与老街入口的地形空间关系，增加游人与孩童的互动。

Two strategies are adopted for the revival of old street:
first, reconstruction and improvement of the quality of overall public space of the village, including the reshape of the topographical spatial relationship between the kindergarten and the entrance of the village, so as to increase the interaction between visitors and children.

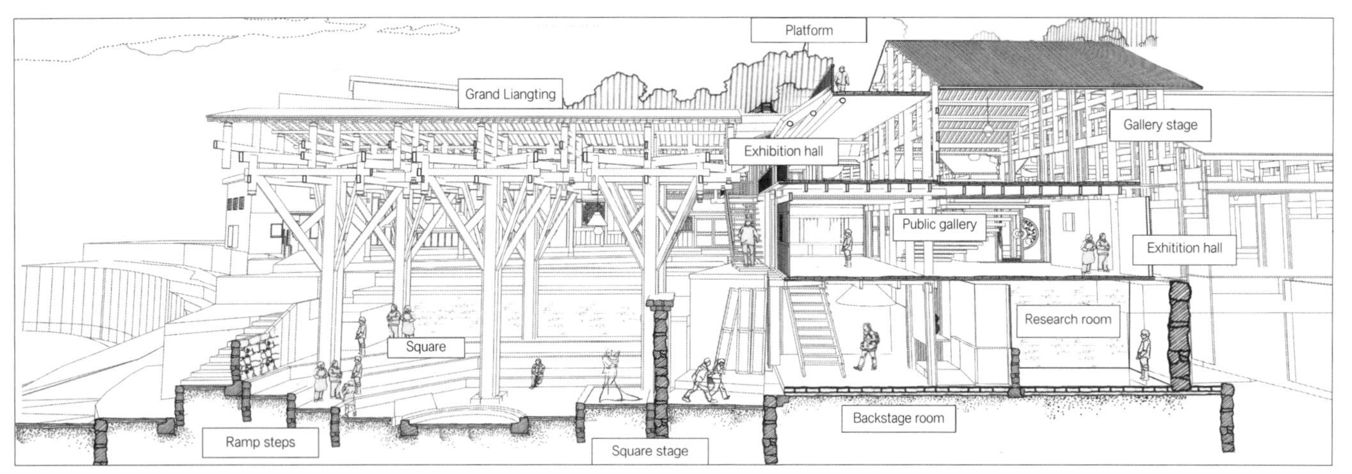

剖透视
Sectional perspective

土家大凉亭村民活动场景
Scenario of villagers activities in Tujia Grand Liangting

土家大凉亭游客活动场景
Scenario of tourists activities in Tujia Grand Liangting

二是单体建筑层面改造。我们拆除了老街中一栋新砖房，以土家吊脚楼特征的木构建筑为原型建造一个凉亭，以恢复老街的传统肌理。大凉亭为村民和游客提供一个集日常集会、民俗节庆、文旅活动、校外课堂等功能于一体的多功能空间。

Secondly, the archtiectural reconstruction was conducted. We demolished one brick house and built an open pavilion which originated from the archtype of Tujia stilted building to recover the traditional texture of the old street. The Grand Liangting provided a multi-functional space for villagers and visitors, forming a public site for daily gatherings, folk festivals, cultural and tourism activities, and off-campus classrooms.

● 西兰卡普工作坊

Silankarp workshop

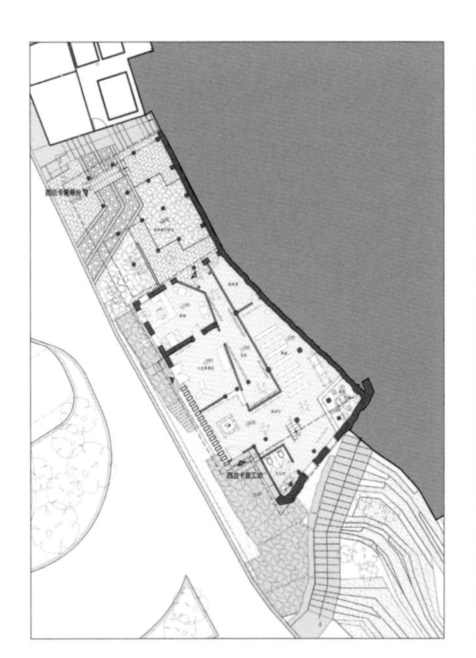

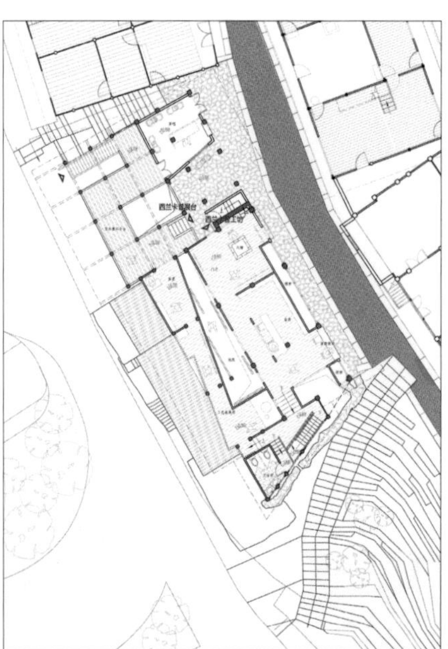

左：西兰卡普工坊底层平面
Left: Ground floor of the Silankarp workshop
中：西兰卡普工坊一层平面
Middle: First floor of the Silankarp workshop
右：西兰卡普工坊模型
Right: Model of the Silankarp workshop
下：西兰卡普工坊剖面空间
Below: Sectional space of the Silankarp workshop

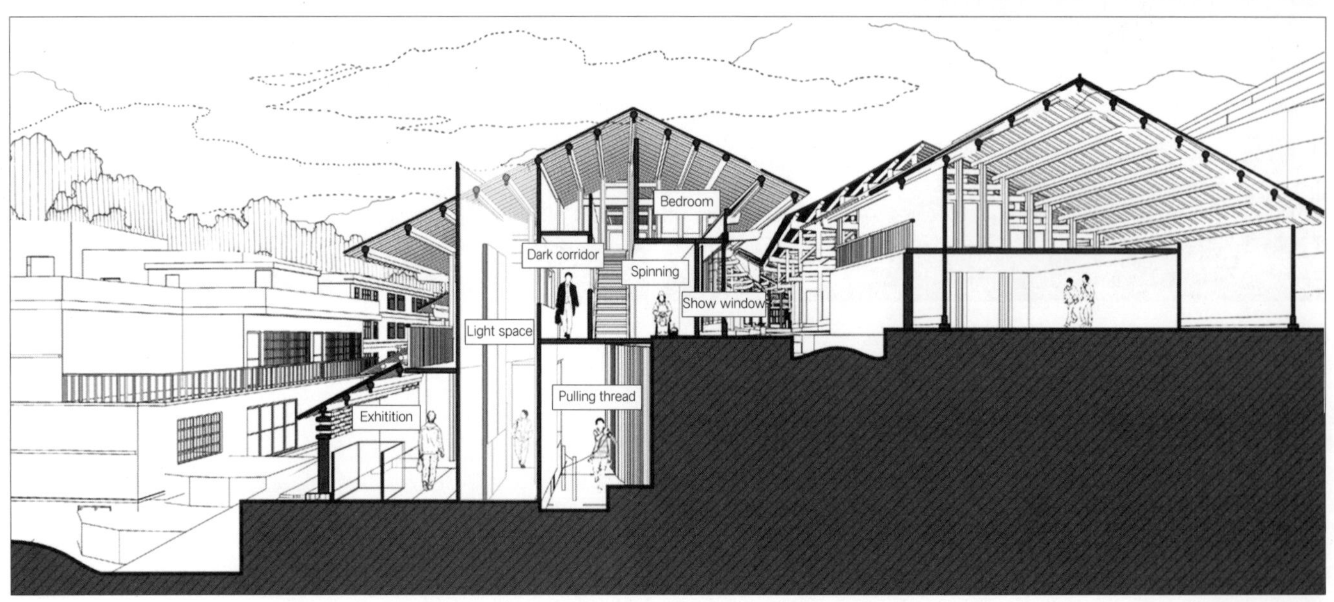

西兰卡普工坊效果图 Rendering of the Silankarp workshop

村口的一栋传统吊脚楼被改造为西兰卡普织锦工坊，将土家族瑰宝西兰卡普的工艺流程与创意产品作为展示主题。设计着重运用光线营造空间氛围，通过置入光井与室外栈台，使得织锦在光影变幻中展现独特魅力，渲染出具有土家民艺特色的空间氛围。

A traditional stilted building at the entrance of the village has been transformed into a Silankarp brocade workshop, showcasing the craftsmanship and creative products of the Tujia treasure Silankarp.The design emphasizes the use of light to create a spatial atmosphere. By placing light wells and outdoor stacks, the brocade exhibits unique charm in the changing light and shadow, rendering a spatial atmosphere with Tujia folk art characteristics.

西兰卡普工坊室内 Interior of the Silankarp workshop

Group 3/SEU 第三组 东南大学

3.3 土家泛文化民宿村设计

Tujia Pan-culture Homestay Village Design

学生：陈韵玄、沈洁、续爽、戈世钊
Students: Chen Yunxuan, Shen Jie, Xu Shuang, Ge Shizhao

指导老师：张彤、王川
Tutors: Zhang Tong, Wang Chuan

在两河口土家泛博物馆旅游开发的背景下，基于老街自身的区位，赋予其文化主题民宿村的功能，既是对当地旅游服务设施的完善，也给沉睡的村庄重新注入活力。恩施土家族苗族自治州土家族文化资源丰富，老街所在的盐道曾是土家文化的传播路线。设计以老街保存完好的空间形态作为物质载体，对当地丰富的土家文化进行转译再现，选取了多种代表性的土家特色文化，提取元素进行转化和空间表达，而非简单的符号呈现。老街北部入口处两栋建筑被用作公共设施，用于日间活动的茶吧和晚间活动的土家酒馆；其余建筑均作为承载一种特定土家文化的主题民宿，选取其中 4 栋建筑进行单体设计，分别是竹编主题民宿、农耕文化主题民宿、傩戏主题民宿，以及有面向街道和室内两个高低不同舞台的土家音乐酒吧。

In the context of the tourism development of the Tujia Pan-museum in Lianghekou, based on the location of the old street itself, endowing it with the function of a cultural themed homestay village not only improves the local tourism service facilities, but also injects vitality into the dormant village. Enshi Tujia and Miao Autonomous Prefecture is rich in Tujia cultural resources. The Salt Road, where the old street is located, was once the transmission route of Tujia culture. Taking its well-preserved spatial form as the material carrier, the abundant local Tujia culture is translated and reproduced. The design selects a variety of representative Tujia characteristic cultures and extracts elements for transformation and spatial expression instead of simple symbol presentation. The two buildings at the north entrance of the old street serve as public facilities, serving as a tea bar for daytime activities and a Tujia tavern for nocturnal entertainment. The rest of the buildings take symbols of a specific Tujia culture respectively. Four buildings are selected for individual design, namely bamboo themed, cultivation culture themed, Nuo Opera themed homestay, and Tujia music bar with two different stages facing the street and indoors.

● 土家傩戏主题民宿
Tujia Nuo Opera themed homestay

傩戏是土家族神秘的传统民俗表演艺术，以傩戏为主题的民宿设计利用了传统民居中阴翳的光影氛围，提供了特色的住宿体验。

Nuo Opera is a mysterious traditional folk performing art of Tujia people. The hotel design with the Nuo Opera theme makes use of the shadow atmosphere in traditional houses to provide a distinctive accommodation experience.

右四图：土家傩戏主题民宿效果图
Right four figures: Renderings of Tujia Nuo Opera themed homestay

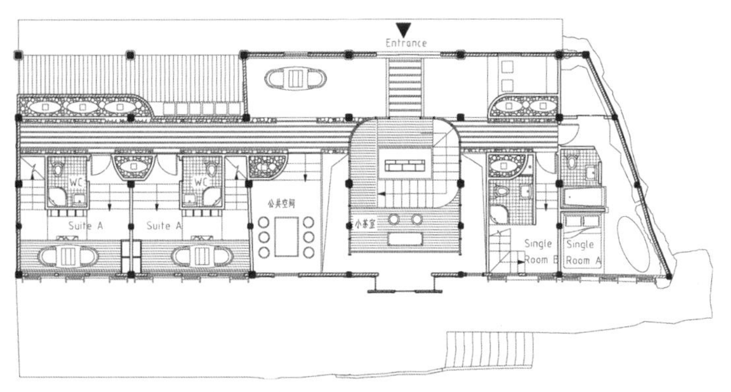

土家傩戏主题民宿一层平面 First floor plan of Tujia Nuo Opera themed homestay

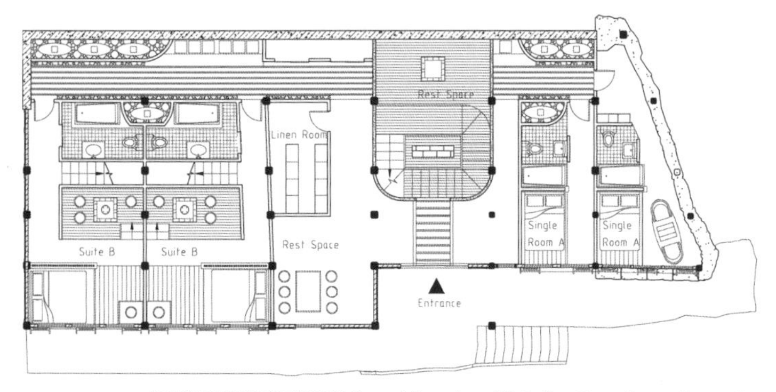

土家傩戏主题民宿二层平面 Second floor plan of Tujia Nuo Opera themed homestay

空间组织流线及功能分区
Spatial organization streamline and functional zoning

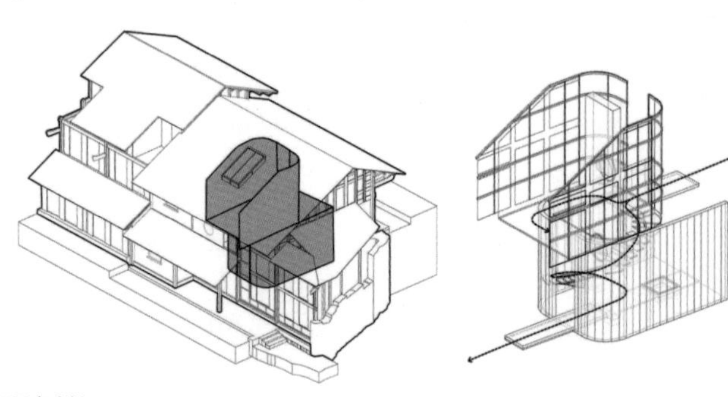

核心空间要素分析
Core spatial elements analysis

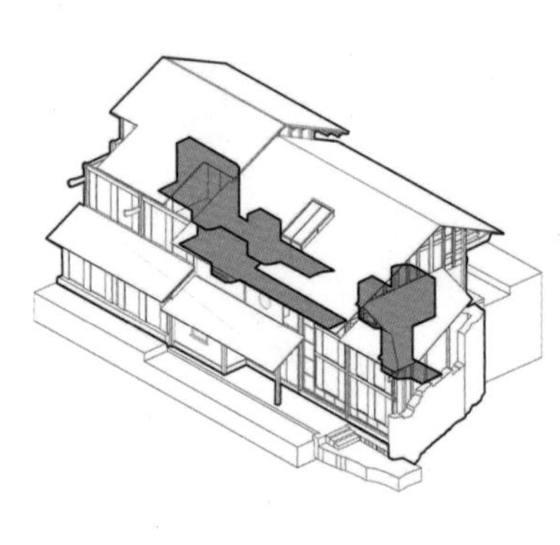

展示空间位置分析
Exhibition space distribution analysis

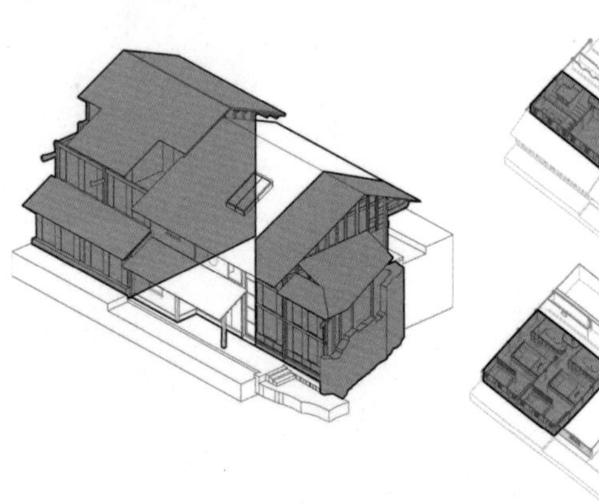

民宿套间类型
Suite types in homestay

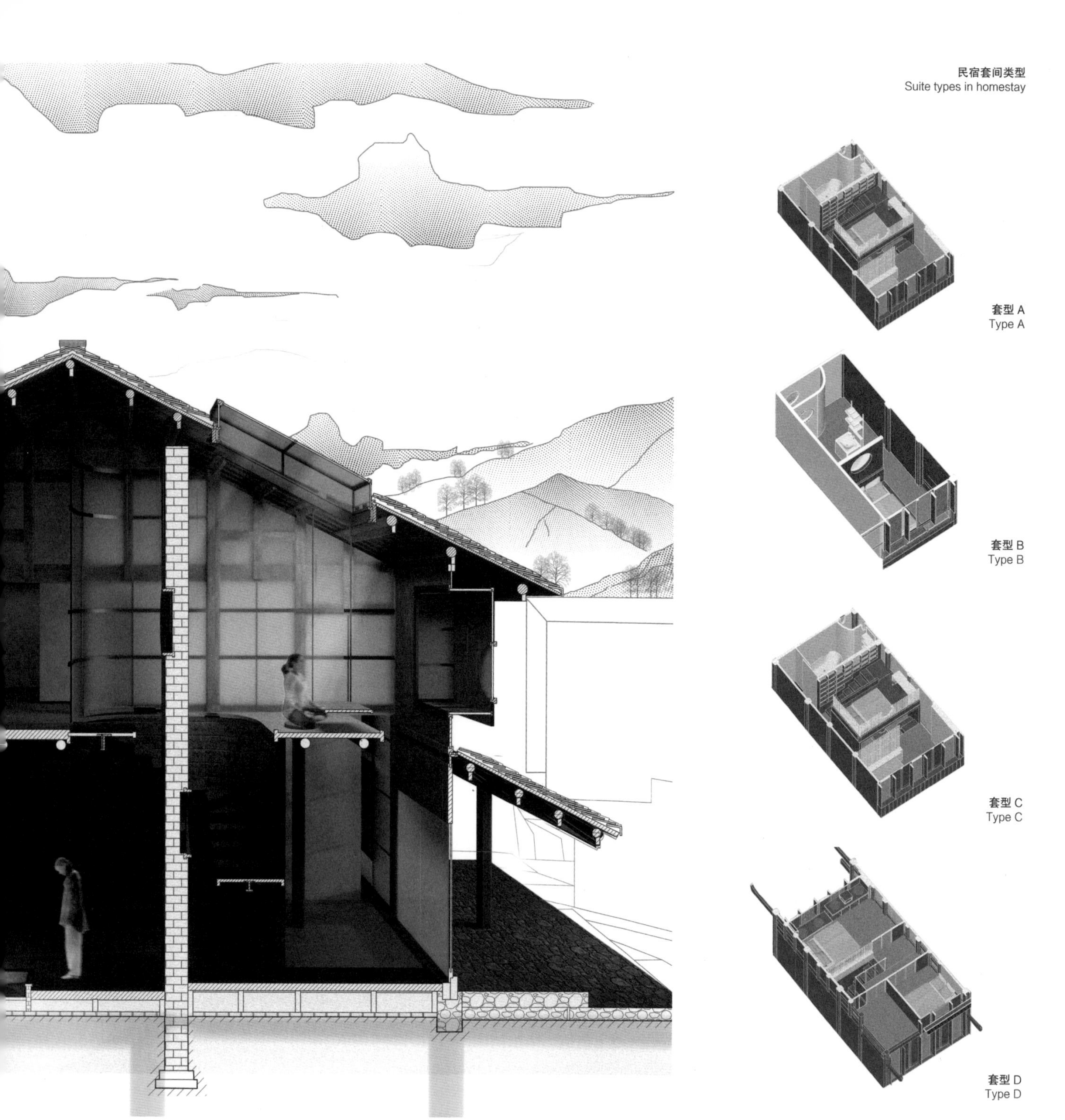

套型 A
Type A

套型 B
Type B

套型 C
Type C

套型 D
Type D

● 土家农耕文化主题民宿

Tujia cultivation culture themed homestay

农耕文化主题民宿效果图 Rendering of the cultivation culture themed homestay

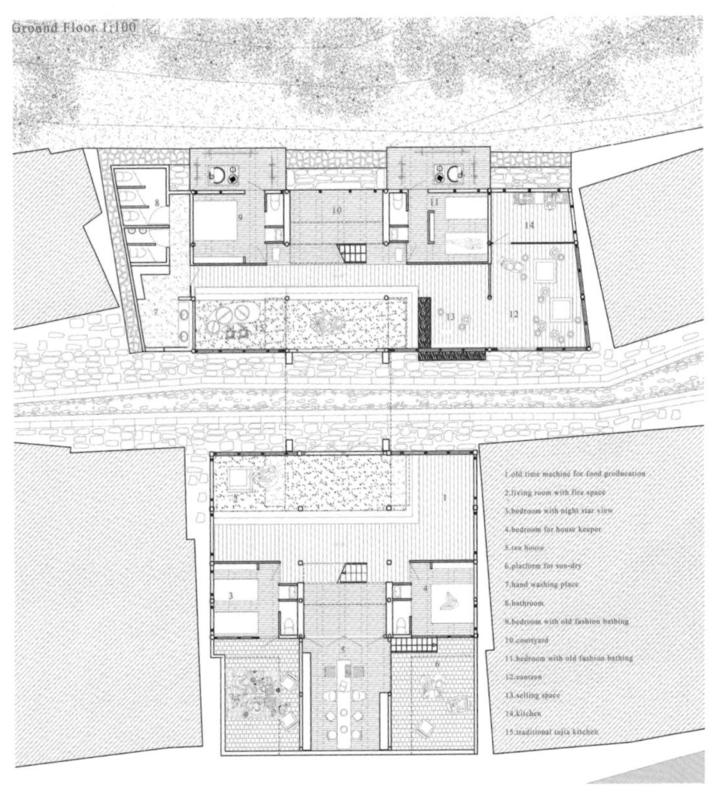

农耕文化主题民宿一层平面 First floor of the cultivation culture themed homestay

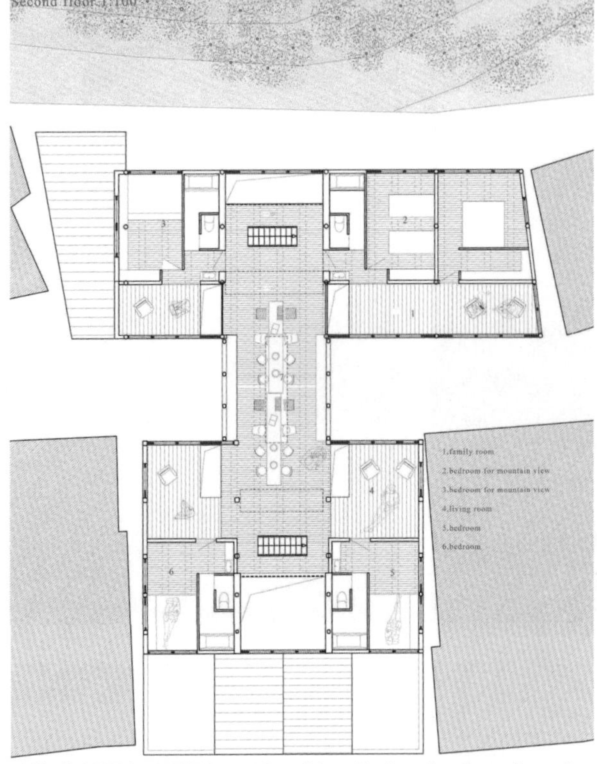

农耕文化主题民宿二层平面 Second floor of the cultivation culture themed homestay

农耕文化主题民宿室内 Interior of the cultivation culture themed homestay

土家族以山林稻作农耕为主要经济生产方式。因此，本方案将土家族农耕文化体验作为核心，通过一系列生活场景来还原传统稻作席居生活。

Tujia people lead a rice farming lifestyle as their main economic production. Therefore, this design takes the Tujia cultivation culture experience as the theme and represents the traditional floor-seating experience through a series of life scenes.

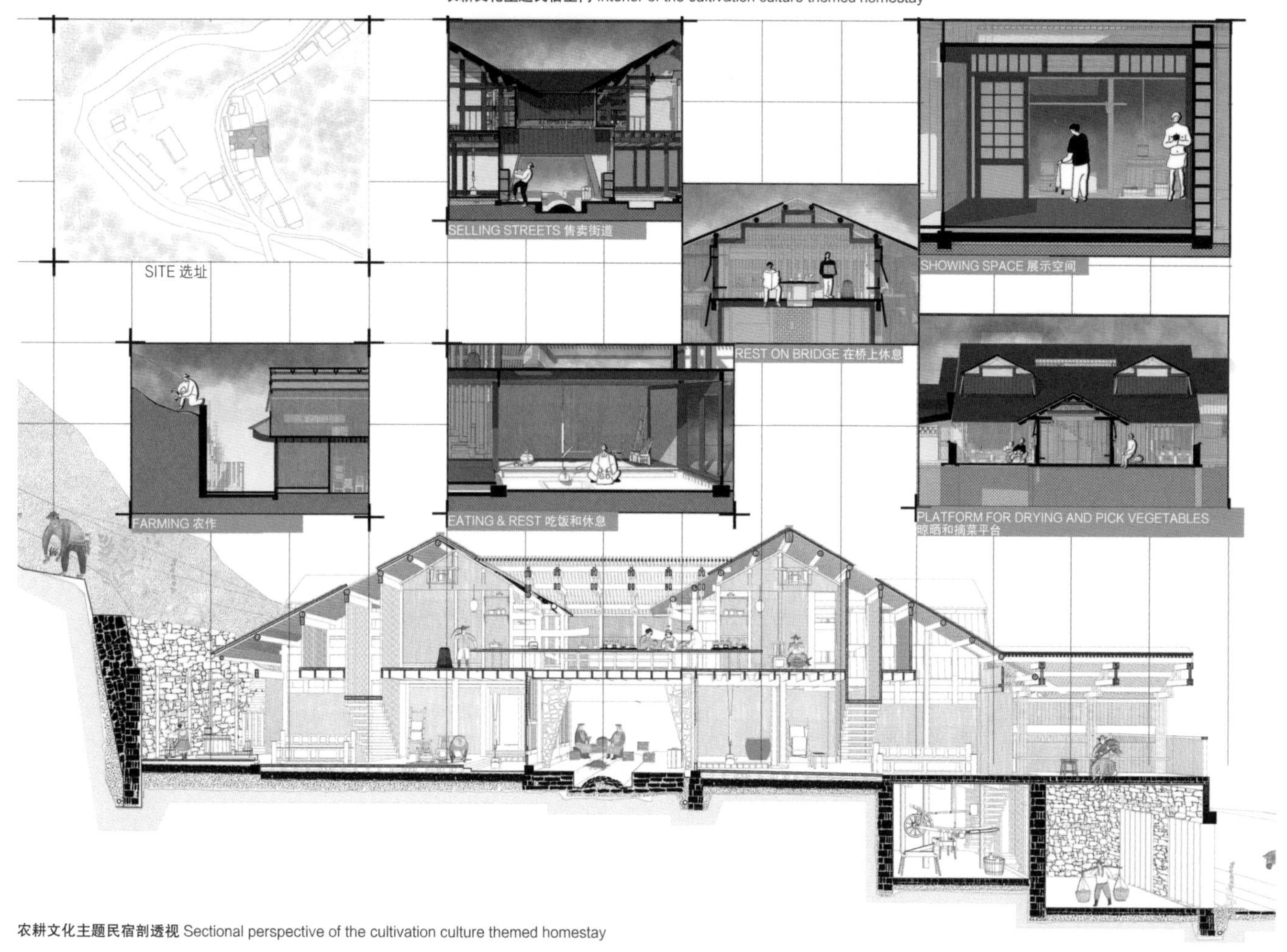

农耕文化主题民宿剖透视 Sectional perspective of the cultivation culture themed homestay

Group 4/HUST 第四组 华中科技大学

3.4 两河集事
Story Market

学生：魏迪、赵蕊、张师维、王沛泽、彭欣怡
Students: Wei Di, Zhao Rui, Zhang Shiwei, Wang Peize, Peng Xinyi

指导老师：李保峰、汤诗旷
Tutors: Li Baofeng, Tang Shikuang

以“两河集事”为主题，一方面表明设计赋予两河口村新的集市功能，另一方面体现了希望更多故事在此发生的愿景。两河口村面临的问题体现为“拆”与“留”的矛盾，我们选择保留老街但转变其功能来面对这一问题，“集事”成为解决问题的突破口。我们通过对聚落、街道、建筑以及行为模式四个方面进行分析评估，提取了老街的七个原型进行拆分与重组。老街一层作为集市，满足当地居民日常购物与赶集的需求，同时也为游客提供文化体验与购物的场所。设计拆除一层部分墙体，保留火塘空间并进行功能转译。二层则作为盐道历史展览空间，植入廊道串联观展流线。除此之外选择三处节点，分别定义为摆手坝、拾忆亭和思阅馆，植入全新功能，为居民与游客提供服务。我们利用两河口老街，尝试创造一个匀质空间，一个可以提供各种事件发生的场所，在这里可能会上演居民与游客的故事、现代与传统的故事以及民族融合的故事……

The theme of "Story Market ", on the one hand, indicates that the design adds the new market function of Lianghekou Village, on the other hand, reflects the vision of wishing more stories happen here. The problem of Lianghekou Village is the contradiction between "demolition" and "preservation". We choose to retain the old street but transform its function to confront this problem, and the story market become the breakthrough to solve the problem. Through the analysis and evaluation of settlements, streets, buildings and behavior patterns, seven prototypes of old streets are extracted, which are divided and reorganized. The first floor of the old street is used as a market to meet the daily shopping and marketing needs of local residents. Meanwhile, it provides a place for tourists to experience local culture and shop. The design removes parts of the wall on the first floor but preserves the fire-pit space and carries out functional translation. The second floor is used as the Salt Road historical exhibition space, and the corridor is implanted to connect the exhibition route. In addition, three nodes are selected, which are respectively defined as waving-hand dam, *shiyi* pavilion and *siyue* hall. New functions are implanted to provide services for residents and tourists. We use the Lianghekou old street to create a public space, where various events can take place, where stories of residents and tourists, modern and traditional stories, and stories of ethnic integration can be staged...

● 土家集市文化
Tujia market culture

Ancient Chuan Salt Road period

After the "Salt and Iron Conference", the mountain people in the northwest of Hubei Province rushed out of the salt area to sell the monopoly of Sichuan salt.

The late Qing Dynasty and the early period of the Anti-Japanese War, two times "Chuanyan Ji Chu", the ancient Salt Road flourished and reached its peak.

After 1949, the ancient Salt Road lost its original function and gradually withdrew from the historical stage.

Emperor Liuche of Han Dynasty
"Salt and Iron Conference"

Emperor Yongzheng of the Qing Dynasty
Tujia clan came under the ruling of Han officials

Revolutionary period

Lianghekou became the resident of the revolutionary regime and established the "Sui'en County Soviet Government".
The main battlefield of suppressing bandits in Lianghekou.

Tourism development

Under the emphasis on the protection and repair of ancient buildings, the site was listed as a provincial-level cultural relics protection unit.
Regional tourism development requirements.

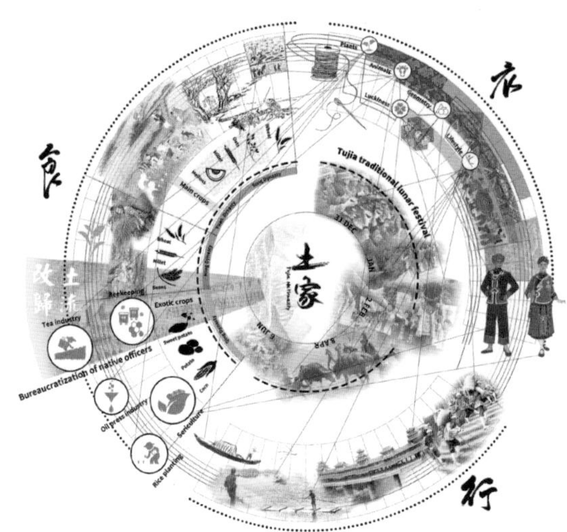

● 老街传统建筑要素分析
Traditional architectural elements in the old street

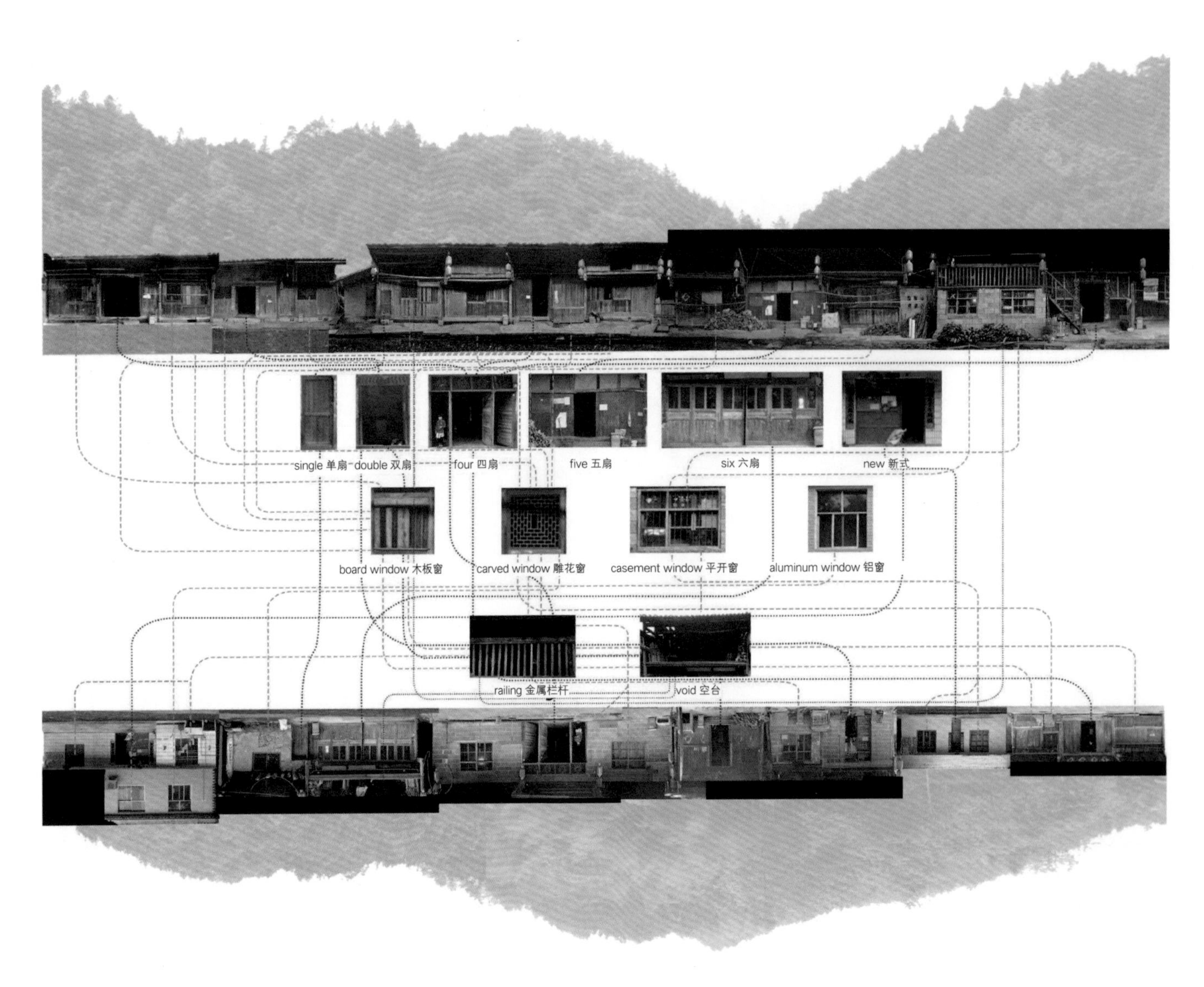

两河口老街既存传统建筑要素分析
Analysis of existing traditional architectural elements in Lianghekou old street

● 两河口集事规划方案
Lianghekou Story Market proposal

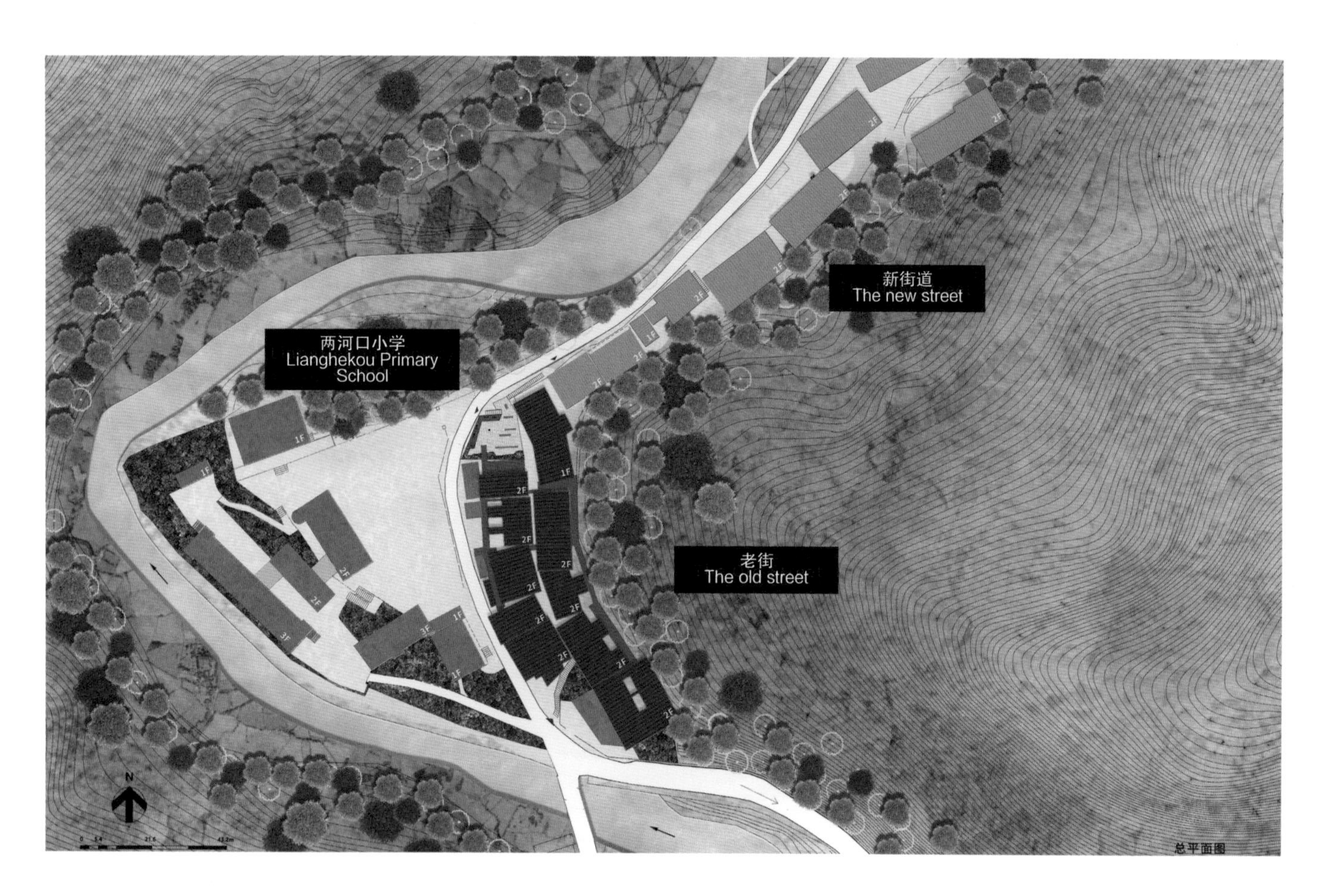

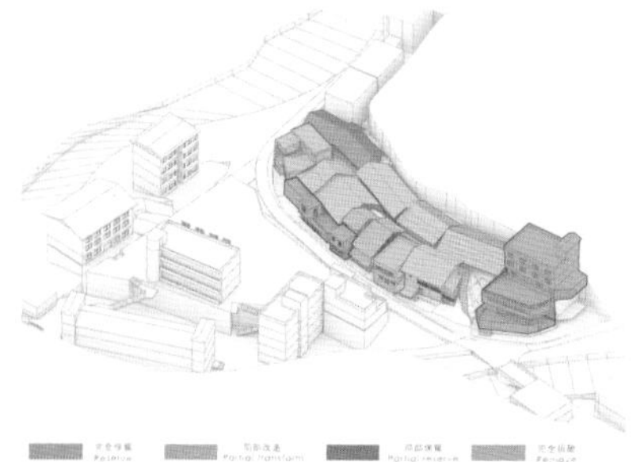

保护原则
Protective principle

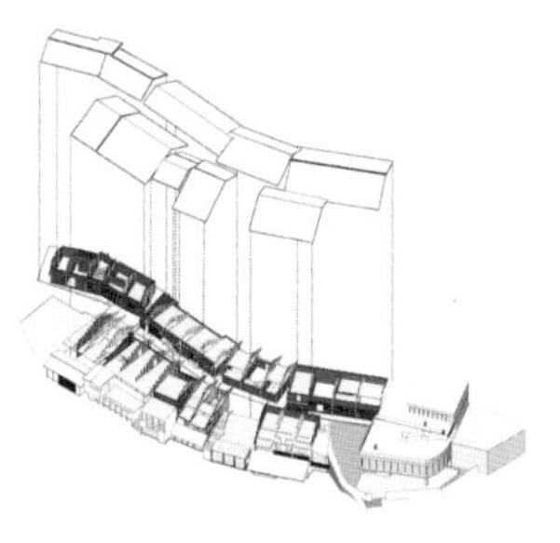

拆除部分要素
Demolished partial elements

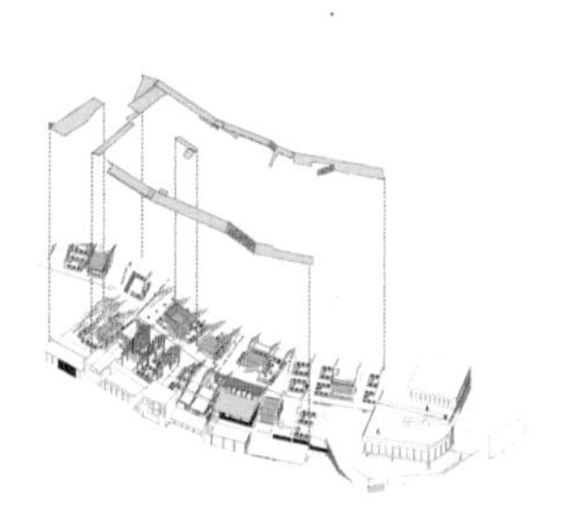

新增部分结构
Added partial new structures

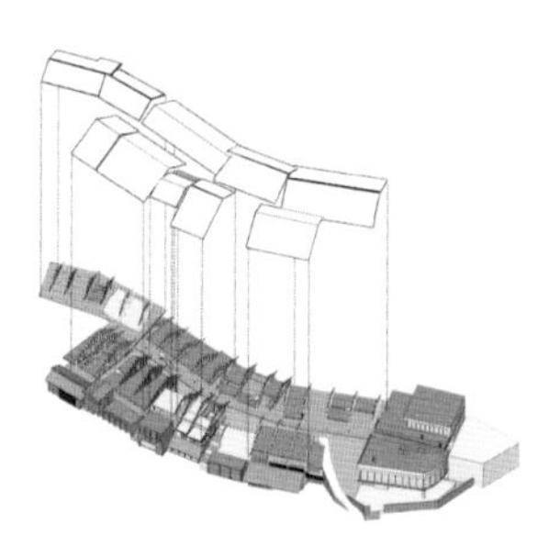

置换部分空间
Transformed partial space

In the past 过去

In natural space 在自然空间

salt trade 食盐贸易

farming 农作

folk dance 民族舞蹈

Tujia festival 土家族节日

In the fire-pit 在火塘里

bacon 熏烤

In the kitchen 在厨房

cooking 做饭

In the hall house 在大堂房内

gathering and ancestor worship 聚会和祭祖

In the alley 在巷子里

rest 休息

At present 现在

In natural space 在自然空间

sightseeing 观光

farming 农作

folk dance 民族舞蹈

Han festival 汉族节日

街道类型 STREET PROTOTYPE

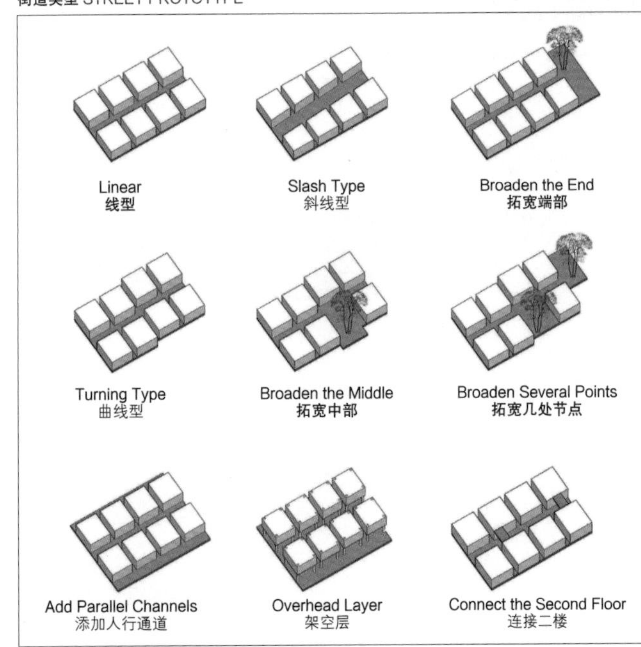

核心功能类型 CORE FUNCTION PROTOTYPE

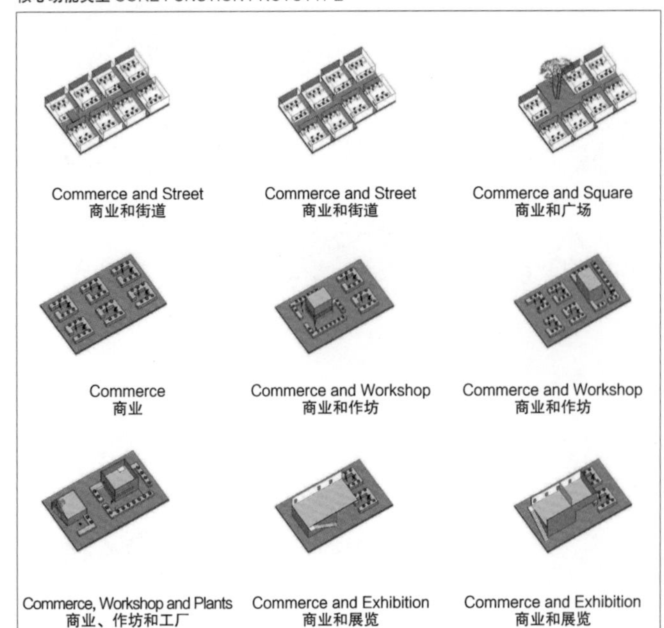

土家集市空间类型与功能分析

Spatial typological and functional analysis of Tujia market

Market 市场

Street 街道

Library 图书馆

Market 市场

Exhibition 展览

Market 市场

Pavilion of memory 记忆馆

Pavilion of memory 记忆馆

Gallery 走廊

Gallery 走廊

两河集市改造后的场景

Scenario after the renovation of Lianghekou story market

Group 5/HUST 第五组 华中科技大学

3.5 言（盐）·趣：两河口盐道古街复兴

Dialogue · Interest: Regeneration of the Salt Road Ancient Street in Lianghekou

学生：王宁、何仕轩、田淑颖（华中科技大学），刘文玉、王长曦（湖北美术学院）
指导老师：李保峰、汤诗旷

Students: Wang Ning, He Shixuan, Tian Shuying (HUST); Liu Wenyu, Wang Changxi (HIFA)
Tutors: Li Baofeng, Tang Shikuang

设计立足于保护、活化古村落，以创造自身“造血”能力为前提，为提升当地居民自身文化认同与延续土家文化而融入土家木工、竹编、西兰卡普等丰富的特色民艺。设计将村落定位为有文化保护、孕育功能的生态博物馆，并将其融入彭家九寨“泛博物馆”体系之内，以达到活化古街的效果。设计转译了“盐”颗粒状的特性，以临街的汇聚各种活动的家庭火塘为核心元素，拆除其外墙，直接将其与古街空间相连，并置入公共的社会活动。火塘作为一系列颗粒状的激活因子，在老街发挥新的作用。此外，为强调土家聚落“山房田水”的空间格局，清除了场地内影响建筑风貌的新建和加建部分，新置入一个社区图书馆，增强其文化属性。同时开辟后山和临水游览路径，中间加以连通，形成网状空间布局。这些活化策略，让人们在盐道古街生态博物馆安居乐业，生成大量的交流言语与丰富有趣且交相辉映的场地空间，实现盐道古街的复兴，故名曰“言·趣”。

This design is based on the protection and regeneration of ancient villages, with the premise of creating their own "hematopoietic" capacity. In order to improve the cultural identity and cognition of local residents about Tujia culture, it integrates rich characteristic folk arts such as Tujia carpentry, bamboo weaving, and Silankarp. The design defines the village as a cultural ecological museum with cultural protection and fertility functions, and integrates it into the Pengjia nine villages' Pan-museum systerm to achieve the aim of regeneration of the ancient street. The design translates the granular characteristics of "salt", with a family fire-pit that gathers various activities on the street as the core element. Its exterior walls are demolished, directly connected to the ancient street space, and incorporated into public social activities. Fire-pits, as a series of granular activating factors, play a new role in the old street. In addition, to emphasize the spatial pattern of "buildings in hills with rivers" in the Tujia settlement, the newly built and added parts that affect the architectural style within the site have been removed, and a community library has been added to enhance its cultural attributes. Meanwhile, the tour paths between the back mountains and the water are opened up, connecting in the middle to form a network spatial layout. These activation strategies, allow people to live and work in the Ecological Museum of the Salt Road ancient street with a large number of sceneries of communicating and interesting space mixed up, achieving the revival of the Salt Road ancient street. Therefore, it is called "dialogue · interest".

● 盐道历史主题的规划设计

Planning and design of the historical theme of the Salt Road

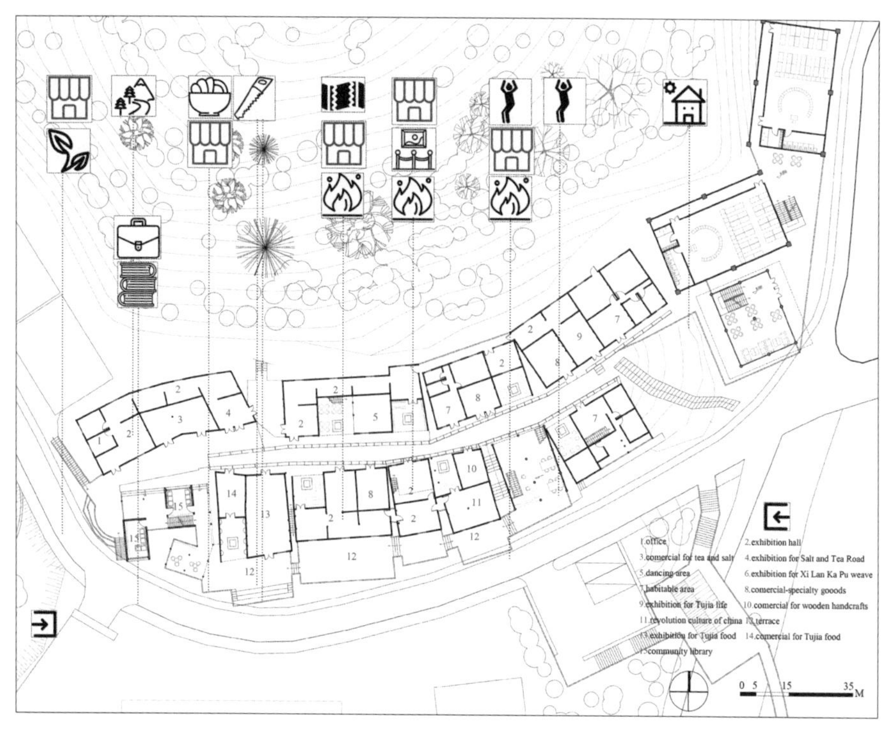

盐道古街的功能更新策划

Functional re-arrangement of the Salt Road ancient street

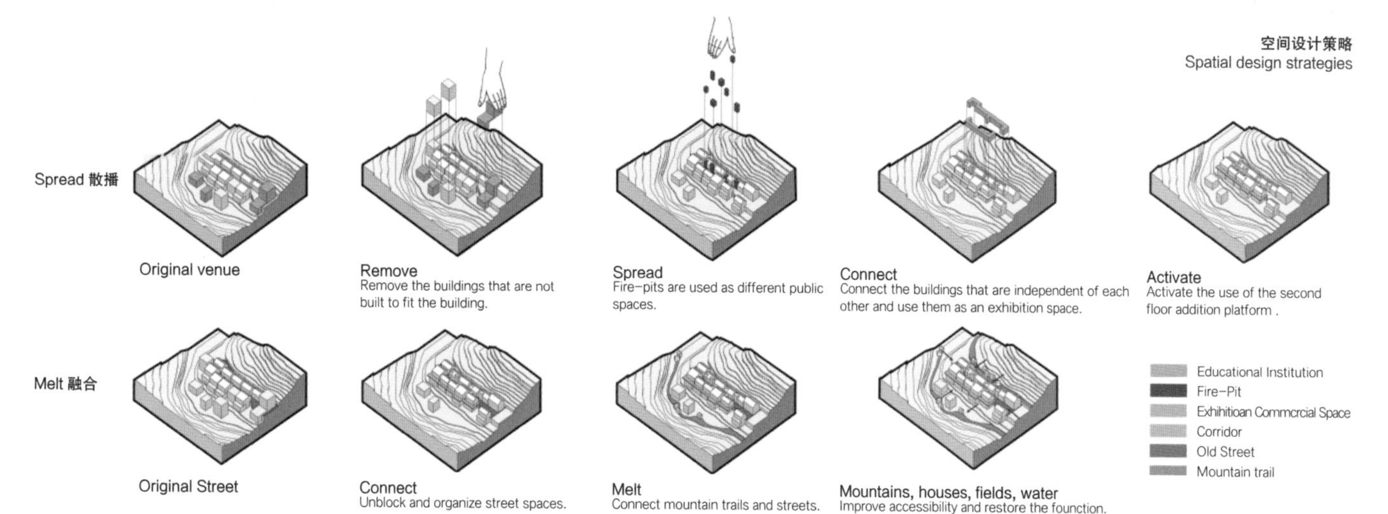

空间设计策略

Spatial design strategies

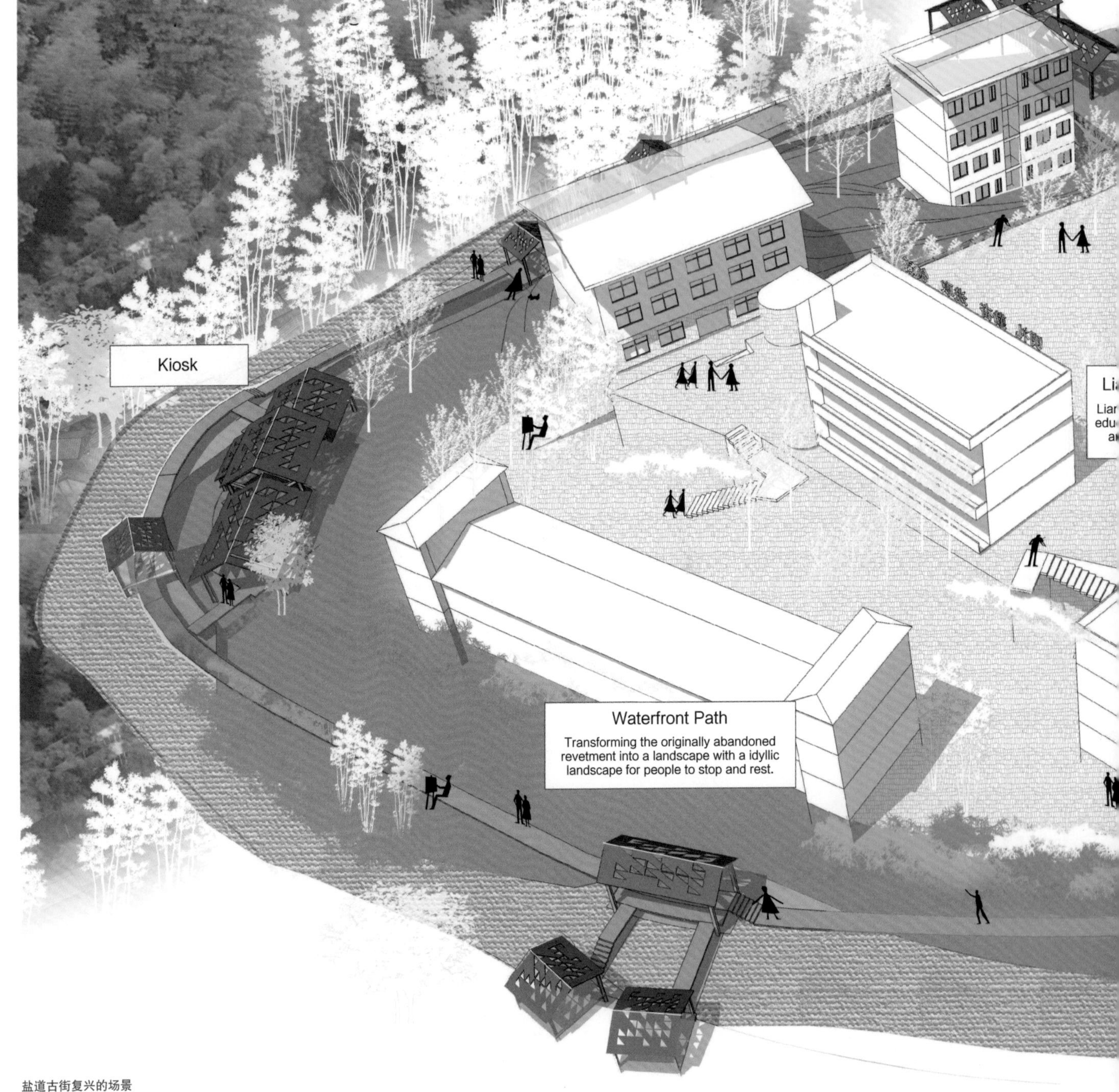

盐道古街复兴的场景
Scenarios of the Salt Road ancient street regeneration

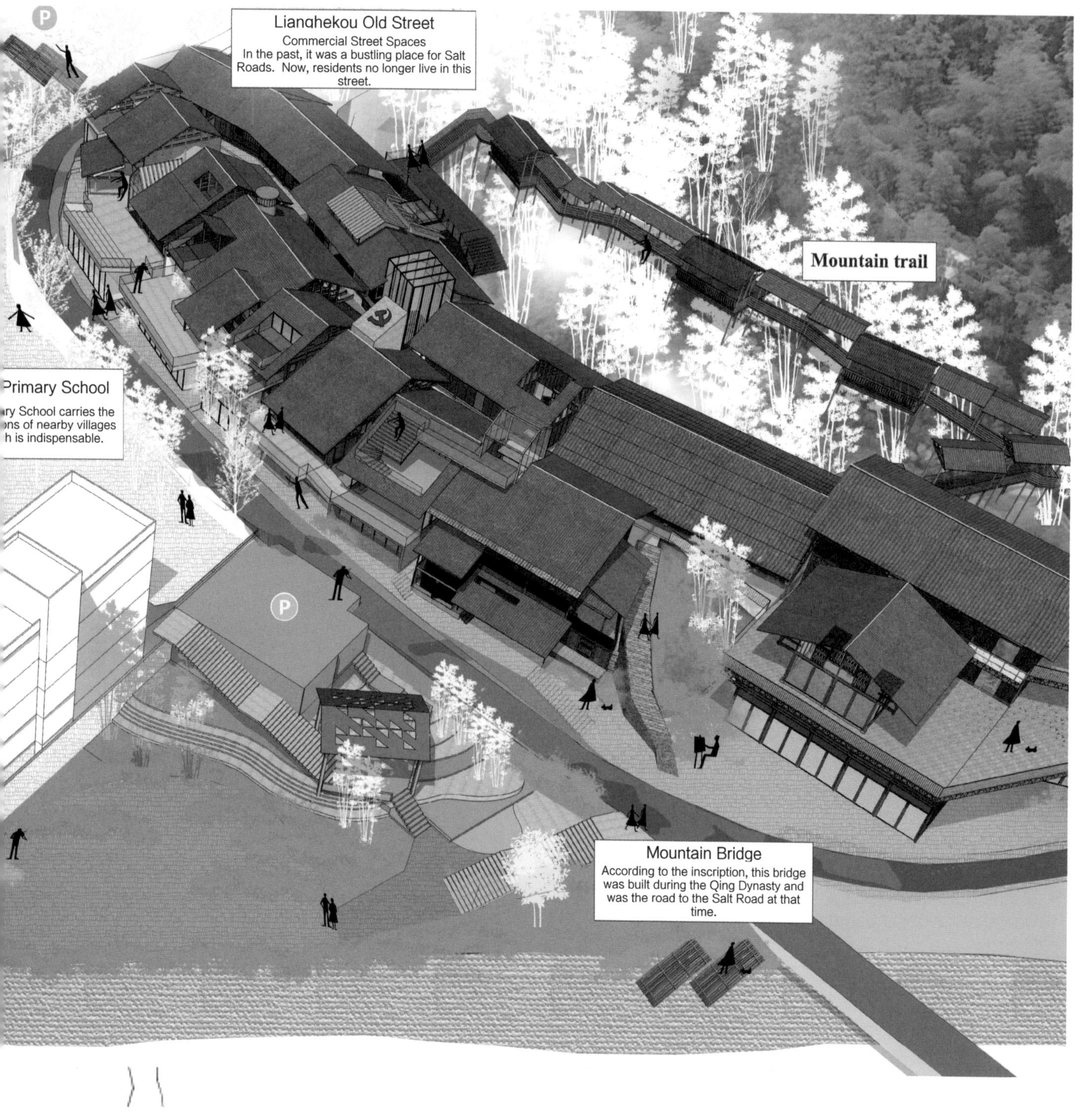
Lianghekou Old Street
Commercial Street Spaces
In the past, it was a bustling place for Salt Roads. Now, residents no longer live in this street.
Mountain trail
Primary School
ry School carries the
ns of nearby villages
h is indispensable.
Mountain Bridge
According to the inscription, this bridge was built during the Qing Dynasty and was the road to the Salt Road at that time.

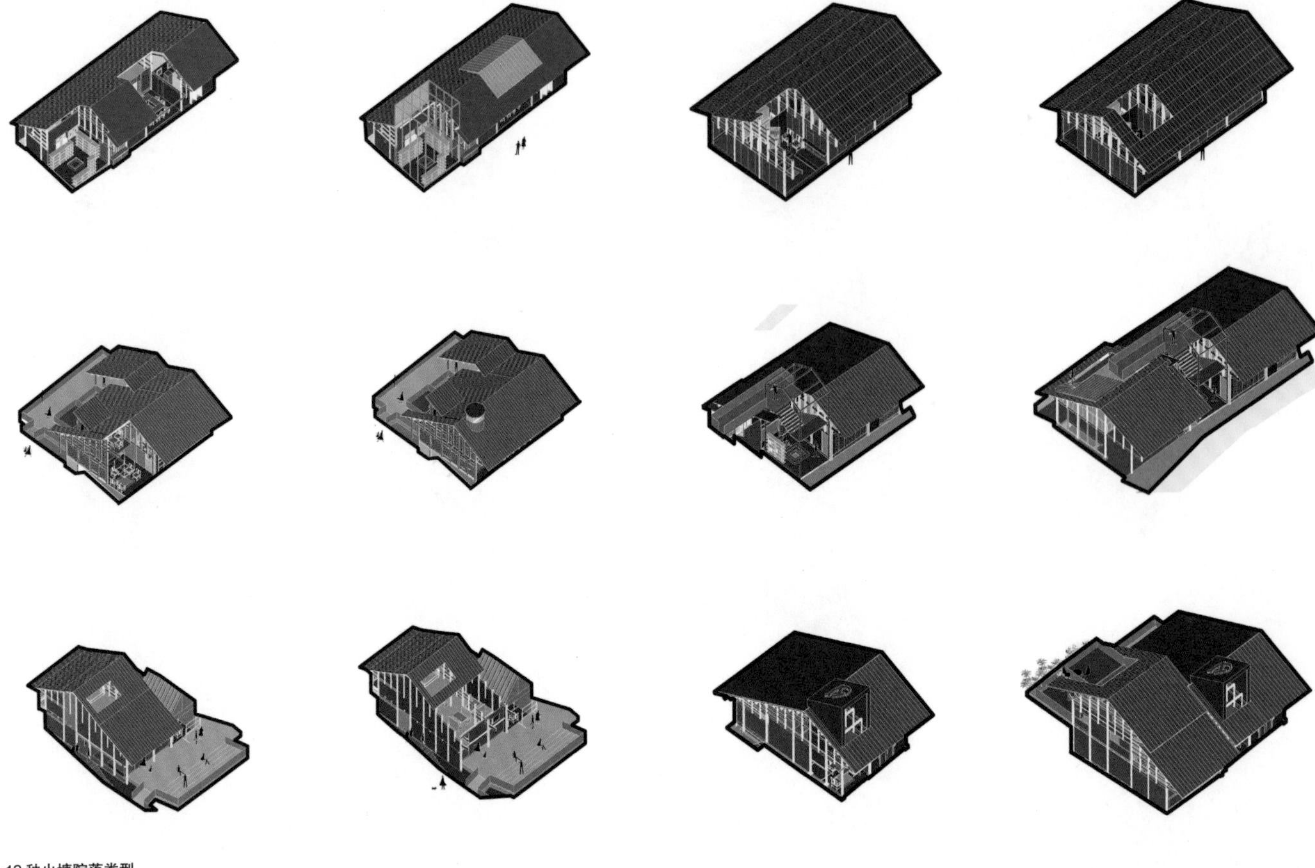

12 种火塘院落类型
12 types of fire-pit courtyards

建筑设计策略
Architectural design strategies

拆除围合墙体的火塘空间，成为激活两河口老街中公共活动空间的重要因子。如此，针灸式的空间改造方式成为传统民居空间组织的重要策略。在改造设计方案中提出 12 种火塘院落的空间构型，为缺乏活力的百年老街提供了全新的活动发生场地。

After the demolition of the enclosuring walls of the fire-pit room, it changes into an important generator of public space in Lianghekou old street. In this way, the acupuncture and moxibustion style space transformation has become an important strategy for the space organization of traditional dwellings. 12 spatial configurations of fire-pit courtyards were proposed in the renovation design plan, providing a new venue for activities in the century old street that lacks vitality.

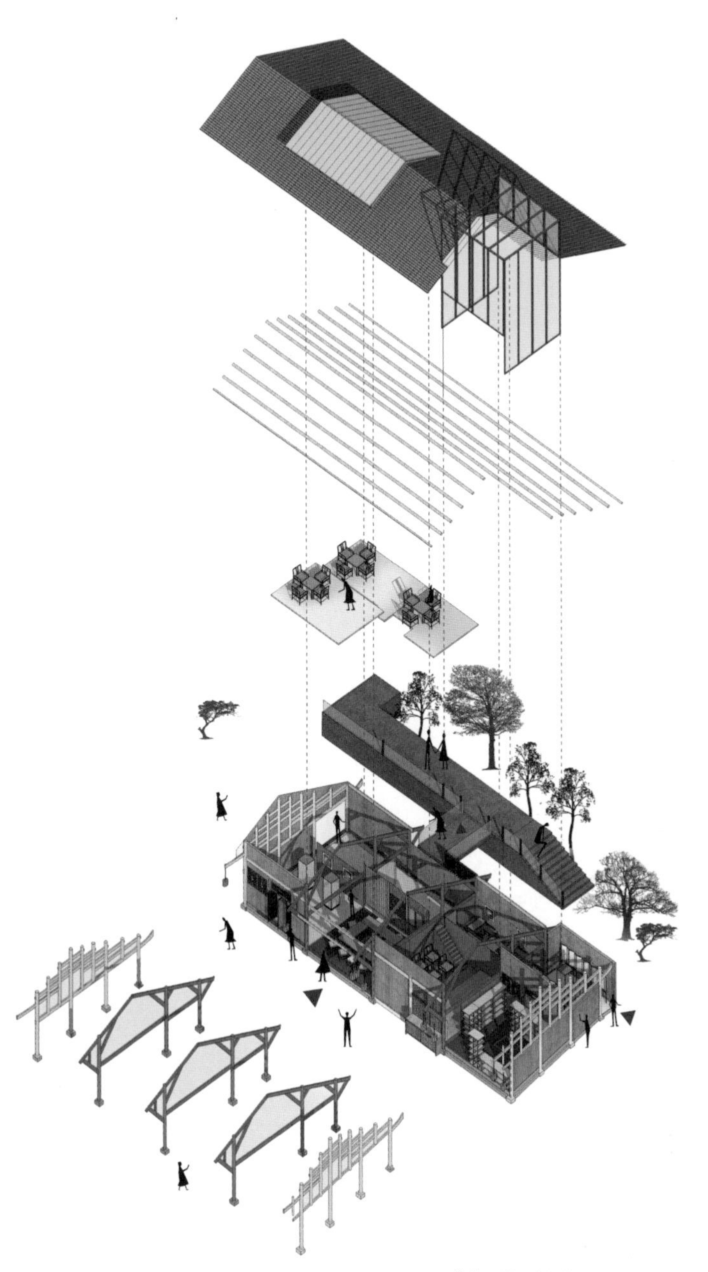

传统民居旧木架与新钢结构的结合
The combination of old timber frames and new steel structures in traditional residence

Group 6/CQU 第六组 重庆大学

3.6 乡村生命体：历史 | 健康 | 和谐

Village Organism: History, Health and Harmony

学生：顾明睿、阳蕊、李晨、吴霜、黄金静、戴连婕
Students: Gu Mingrui, Yang Rui, Li Chen, Wu Shuang, Huang Jinjing, Dai Lianjie

指导老师：褚冬竹、宫聪
Tutors: Chu Dongzhu, Gong Cong

针对村庄现存问题，通过对村民5分钟、10分钟、15分钟活动圈的分析，我们将老街的一期建设定位为村民活动中心。在规划、建筑、景观三方面分别提出“空间处方”，意图用老街这个“乡村生命体”“医治”两河口村的“伤病”。在规划上，我们分析出人口结构不平衡、村庄空置、文化生活匮乏、产业结构单一这四方面主要矛盾，所以我们一方面运用空间句法对场地交通进行重新梳理，另一方面置入茶馆、文具店、体检中心、民宿、图书馆、体验馆等多种功能。在景观上，由于原老街景观单一，于是选取两轴（新街、老街）和三节点（南入口、北入口、后山）进行景观优化。在建筑上，对12栋房子建立档案卡，评估出3栋最具改造潜力的建筑进行改造。古建筑1位于整条老街的端头，设计重新梳理了高差，利用小的错层和楼梯制造更多趣味空间，希望通过历史场景的叠加，让静止的建筑动起来；古建筑2倚靠后山，设计在保留并优化原有村民居住功能的同时，将闲置空间改造为民宿，希望通过平静的介入让建筑功能和谐演变；古建筑3也位于老街的端头，同时面对小学，设计置入老年健康、老年活动、儿童餐饮等主题功能，保留古建筑的整体结构，再现传统手工艺的同时，也为功能新生提供可能。

In view of existing problems in the village, through analysis of the villagers' 5-min, 10-min, 15-min activity circles, we position the first phase of the construction of the old street as a village activity center. We put forward "space prescriptions" in the planning, construction and landscape respectively, expecting that the "village organism" "cures" the "injury" of Lianghekou Village. In terms of planning, we analyzed the main contradictions in four aspects: imbalanced population structure, vacant villages, lack of cultural life, and single industrial structure. Therefore, on the one hand, we used spatial syntax to reorganize the site transportation, and on the other hand, we placed various functions such as tea houses, stationery stores, medical examination centers, homestays, libraries, and experience centers. In terms of landscape, due to the single landscape of the original old street, two axes (new street, old street) and three nodes (south entrance, north entrance, and back mountain) were selected for landscape optimization. In terms of architecture, archive cards were established for 12 buildings, and 3 buildings with the most potential for innovation were evaluated for renovation. Traditional building 1 is located at the end of the whole old street. The design reorganizes the height difference and makes use of small splitter floors and stairs to create more interesting space. It hopes to make the static architecture perform through the overlay of historical scenes. Traditional building 2 relies on the back mountains. While retaining and optimizing the original residential functions of villagers, the design upgrades the spare space into a homestay, hoping to make the building function evolve harmoniously through peaceful intervention. Traditional building 3 is also located at the end of the old street, facing the primary school. The design incorporates theme functions such as elderly health, elderly activities, and children's dining, preserving the overall structure of the ancient building, reproducing traditional handicrafts, and providing possibilities for functional renewal.

● 规划策略
Planning strategies

改造后老街鸟瞰
Overview of the renovated old street

Phase Ⅰ: The main building functions and structural repairs on the street were completed, and the main service personnel of the building were villagers. Reserve about 30% of the building for subsequent development.

Phase Ⅱ: Establish a new health center on one side of the old street. Small shops such as coffee shops and souvenir shops have been added to the old street.

Phase Ⅲ: Develop the area to the bridge. In addition to the existing functions, the function of the reserved land is rationally laid out.

■ 交通规划 Transportation plan ■ 功能分区 Functional divisions ■ 功能分层 The functions of different floors

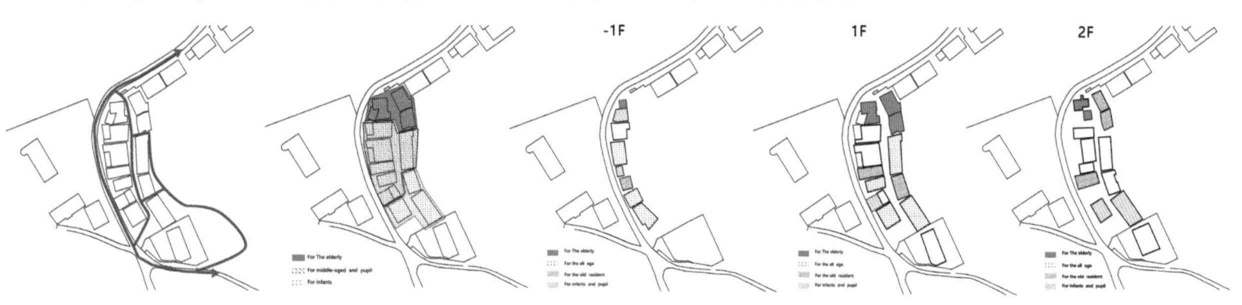

■ 街道空间分析 Street space analysis

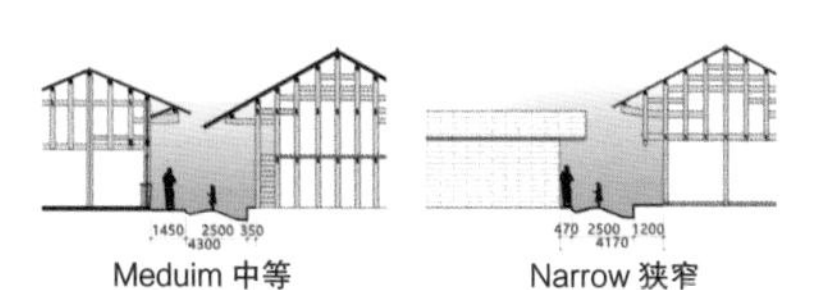

■ 视线分析 Visiability analysis

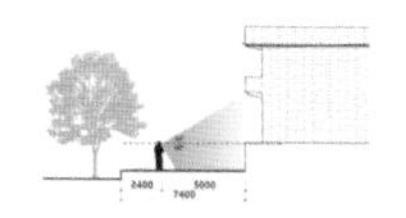

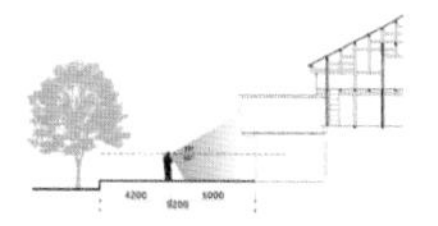

改造设计策略
Strategies of renovate design

● 建筑档案卡
Architectural archive card

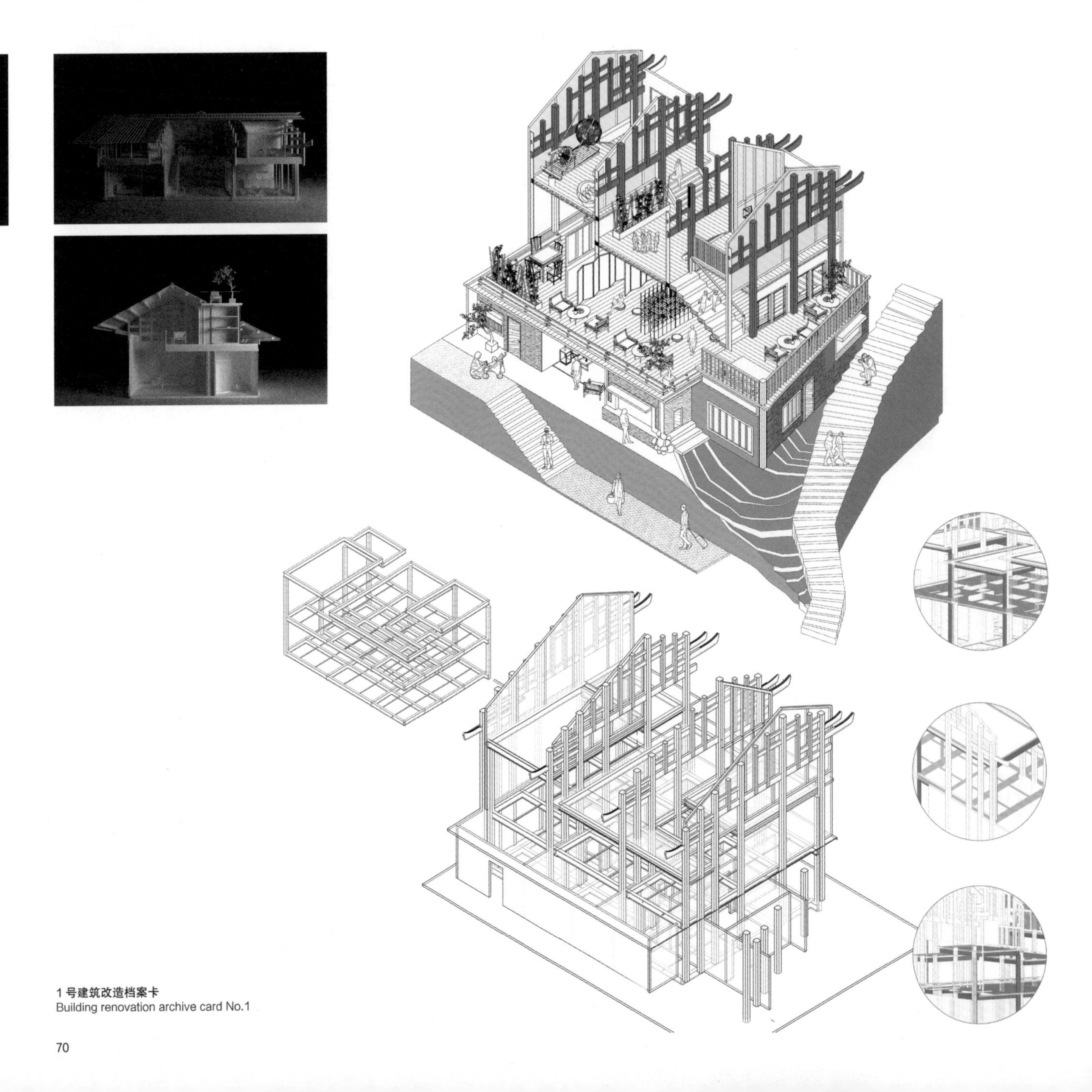

1 号建筑改造档案卡
Building renovation archive card No.1

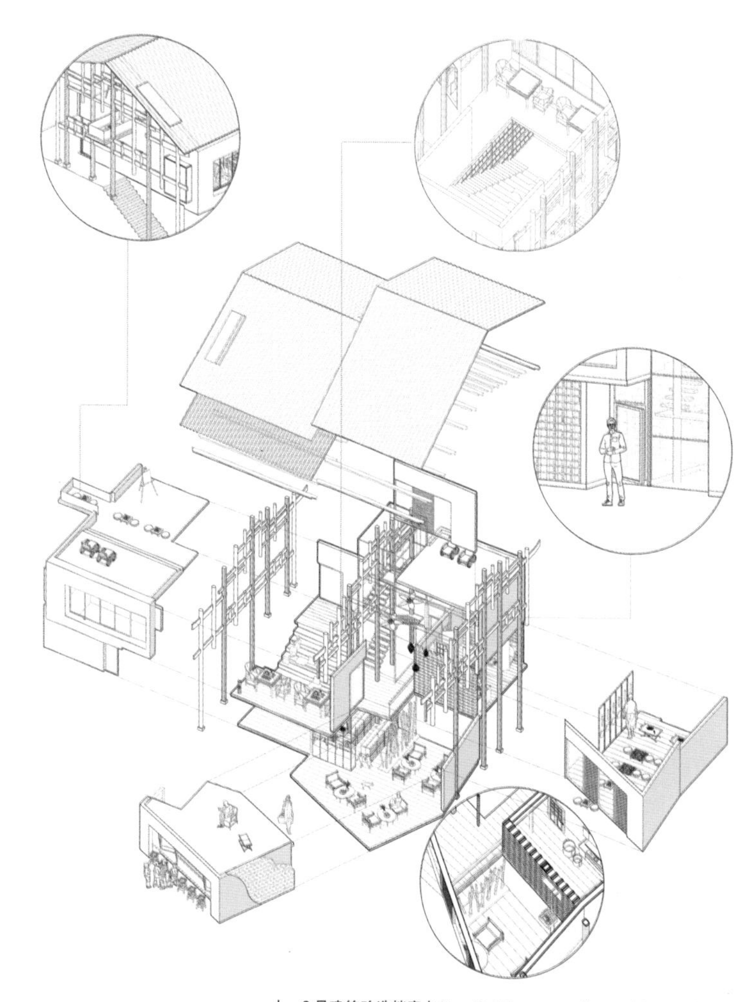

上：2号建筑改造档案卡 Up：Building renovation archive card No.2
下：新建木结构的多重功能 Below: Multiple functions of the newly Built wooden structure

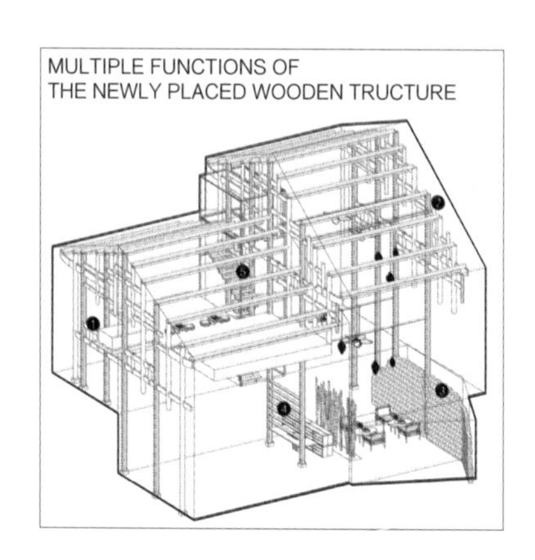

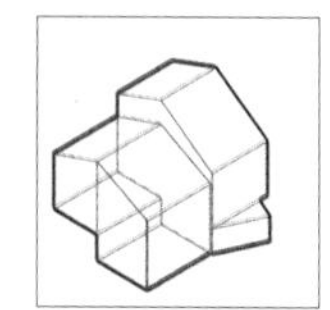
The massing of new house

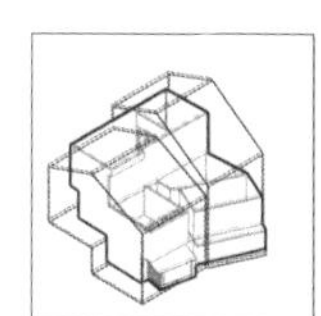
The massing of original house

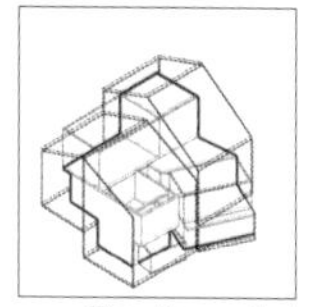
The dismantled

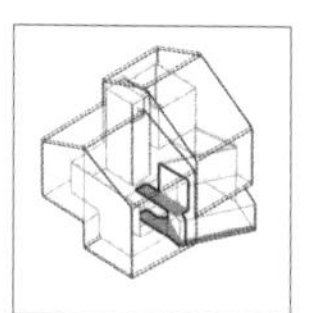
The additions

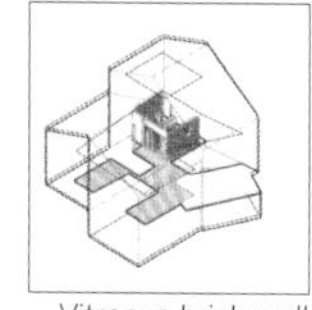
Vitreous brick wall

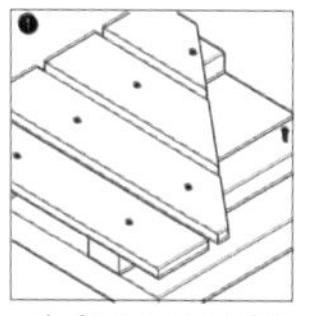
1. As a connection with the roof terrace

2. As a connection with the ceiling keel

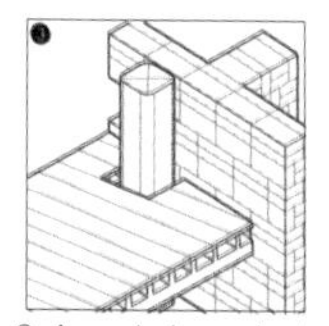
3. As an independent structual system

4. As a component of a kids'climbing ladder

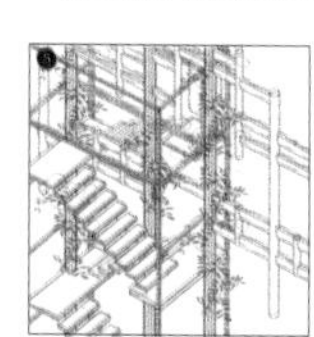
5. As an installation art with hanging plants

Group 7/IUAV 第七组 威尼斯建筑大学

3.7 竹钢
The Green Steel

学生：埃莉萨·切洛、安娜·马尔赛拉
指导老师：阿尔多·艾莫尼诺、恩里科·丰塔纳里、朱塞佩·卡尔达罗拉

Students: Elisa Cielo, Anna Marsella
Tutors: Aldo Aymonino, Enrico Fontanari, Giuseppe Caldarola

“竹钢”项目提出了另一种可行的替代方案：通过改变结构的材料来与传统兼容。利用竹子这种绿色生态材料，结合当代建造技术，形成“绿色钢铁”，展示了具有建设性的技术解决方案，以获得兼具多种功能的内部空间和与传统融合的室外布局。在处理不同层次的保护和创新时，总体外部形象得以保持。通过修补恢复建筑围护结构来实现与周围风貌相合的传统意象。通过改变传统的木结构，并用竹子（一种自然有机的“钢”建造材质）取代它，获得了一种新的适应性内部分布：通过低成本和低技术的干预，产生了更适用的内部空间，保持了传统的兼容室外布局。

"The Green Steel" project suggests a possible alternative way to obtain a generally compatible traditional image by changing the materials used for the structure. It works with bamboo (the green steel, as in the evocative title), showing constructive technical solutions to obtain versatile internal spaces with a traditionally compatible outdoor layout. Dealing with different levels of conservation and innovation, the general external image is maintained. A generally compatible traditional image is set by restoring the building envelope. A new adaptive internal distribution is obtained by changing the traditional wooden structure and substituting it with bamboo (a green construction material like steel): more versatile internal spaces are generated with low-cost and low-tech interventions, maintaining a traditionally compatible outdoor layout.

● 规划策略
Planning strategies

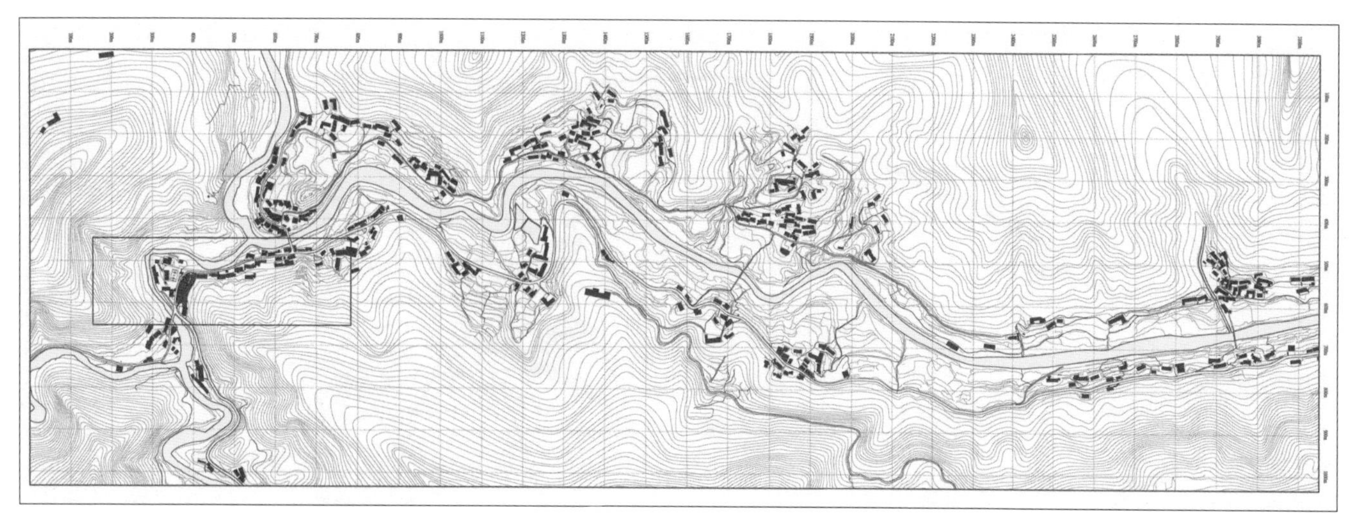

山谷总平面图
Site plan of the valley

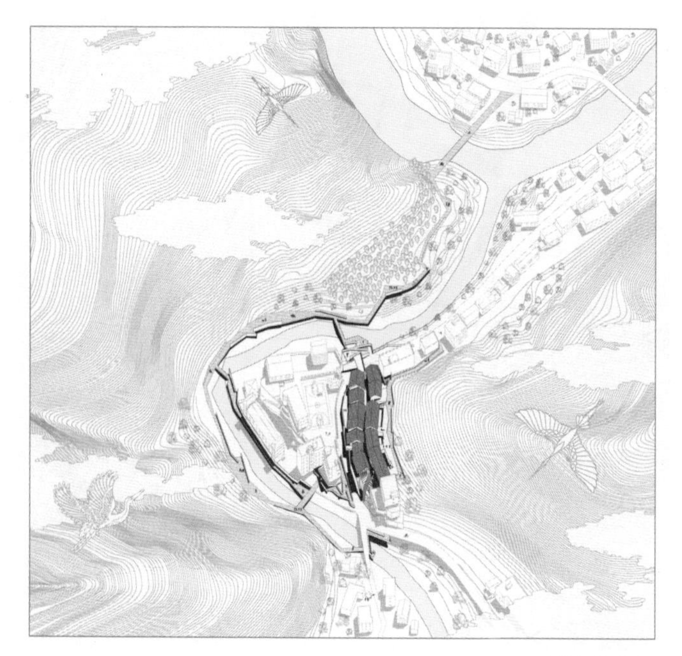

两河口村总平面 1
Masterplan of Lianghekou Village 1

两河口村景观
Lianghekou Village landscape

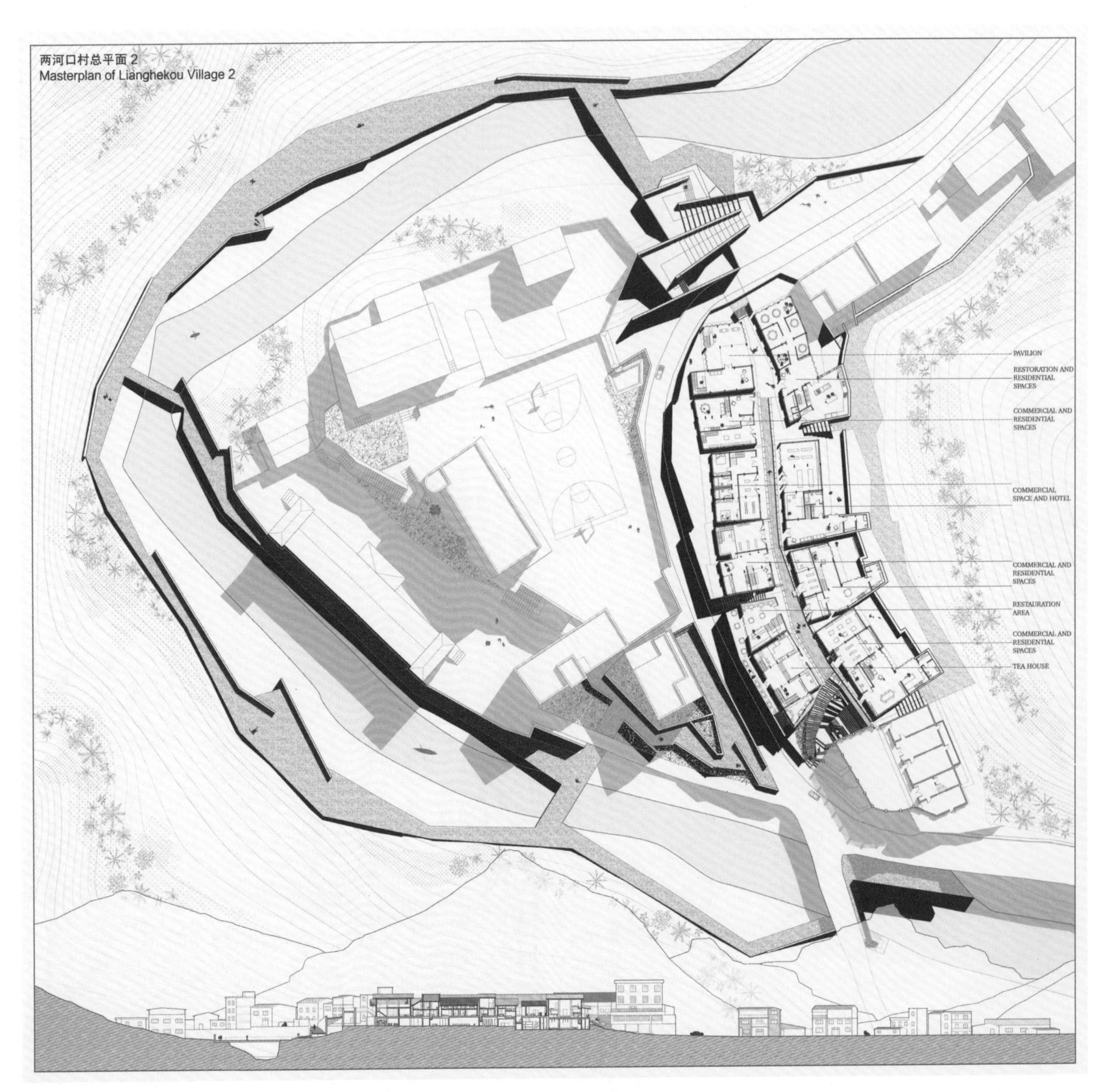
两河口村总平面 2
Masterplan of Lianghekou Village 2
PAVILION
RESTORATION AND RESIDENTIAL SPACES
COMMERCIAL AND RESIDENTIAL SPACES
COMMERCIAL SPACE AND HOTEL
COMMERCIAL AND RESIDENTIAL SPACES
RESTAURATION AREA
COMMERCIAL AND RESIDENTIAL SPACES
TEA HOUSE

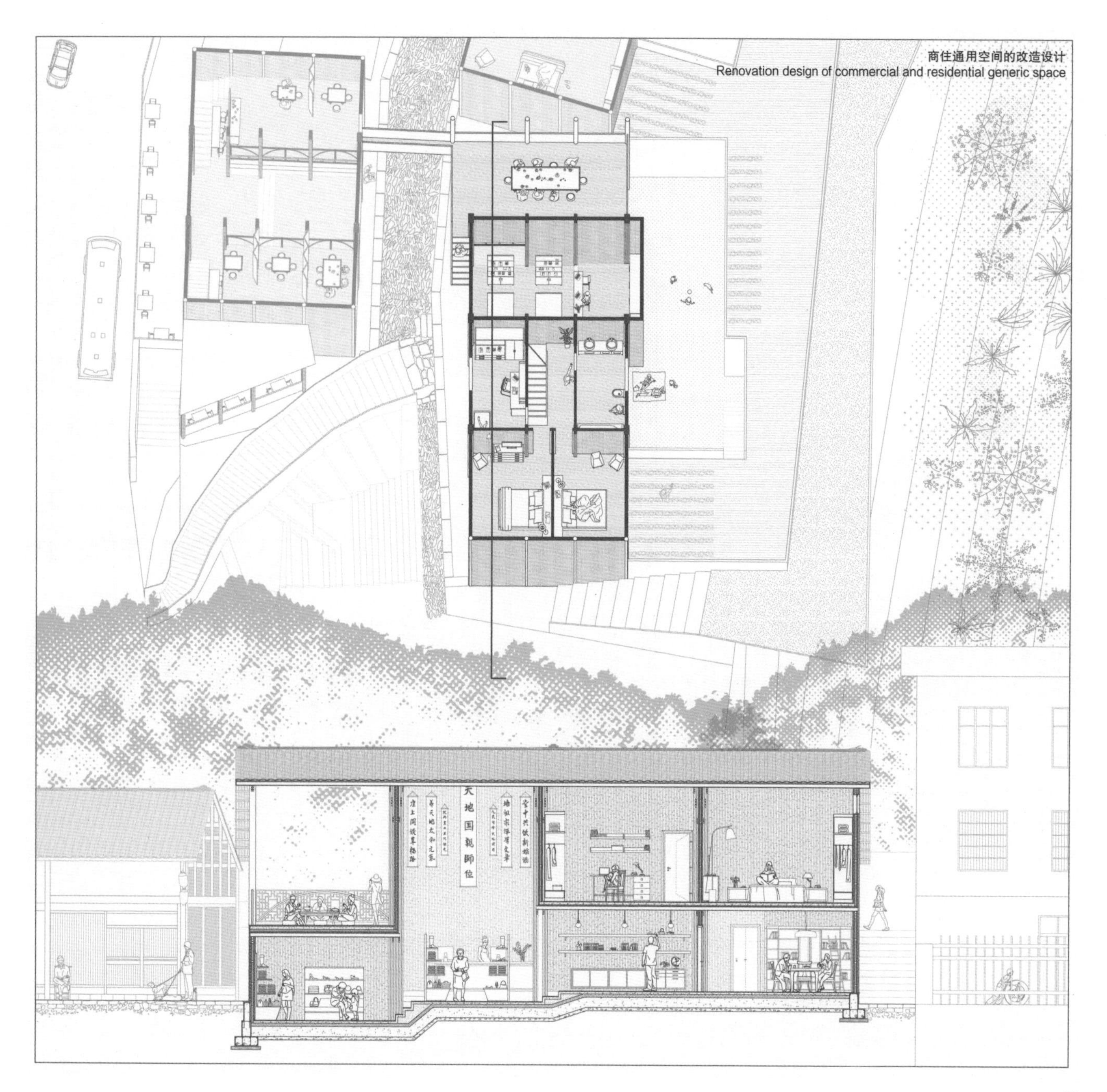

商住通用空间的改造设计
Renovation design of commercial and residential generic space

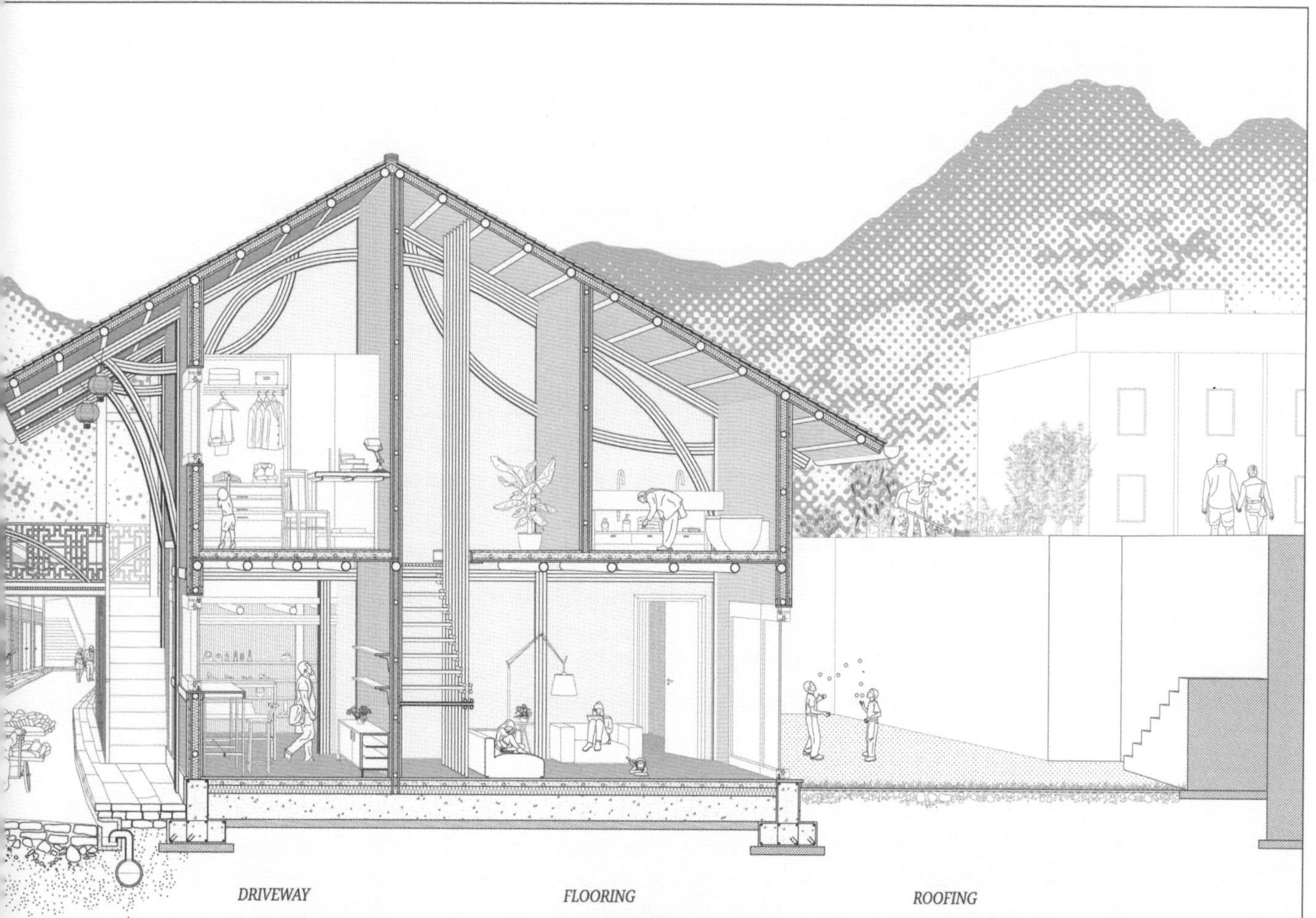

DRIVEWAY

2.5 % slope wear surface
leveling coat
driveway gutter
cockpit
mixed granular well constipated bottom
lean concrete

FLOORING

bamboo flooring 20 mm
impact soundproofing mat 2 mm
installations screed 75 mm
insulator 70 mm
vapour barrier 2 mm
double wood crossed planking 20 mm+20 mm
bamboo joists120mm
bamboo beams 3×5 50 mm

ROOFING

tiles
waterproofing membrane 2 mm
boarding 25 mm
insulator 75 mm
vapour barrier 2 mm
bamboo joists120 mm
boarding 20 mm
bamboo beams 3×5 50 mm
gutter canal

SALT ROAD

coat of well woven pebbles
siphon trap
sewer Φ 100
conduct pipe Φ 300
stabilized fine gravel bottom
draining gravel
drainage pipe
geotextile
ground

FOUNDATIONS

bamboo flooring 20 mm
installations screed 80 mm
insulator 100 mm
slab 300 mm
lean concrete 100 mm
foundation plinth

WALLS

bamboo cladding 25 mm
waterproofing membrane 2 mm
horizontal boarding 25 mm
vertical boarding 25 mm
insulator 80 mm
vertical boarding 25 mm
wooden sliding door with wooden frames

两河口盐道老街剖透视
Sectional perspective of the Salt Road ancient street in Liagnhekou

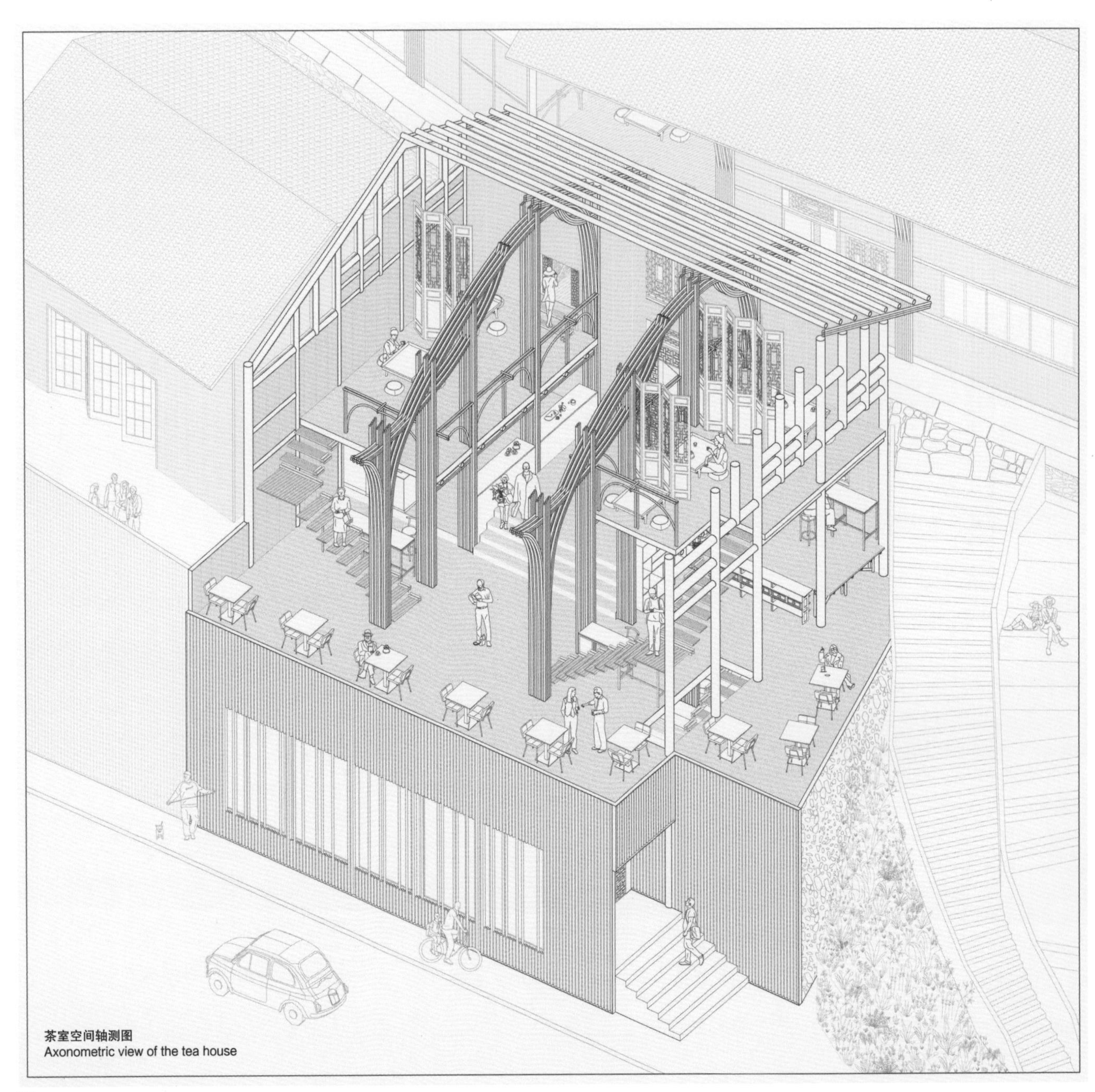

茶室空间轴测图
Axonometric view of the tea house

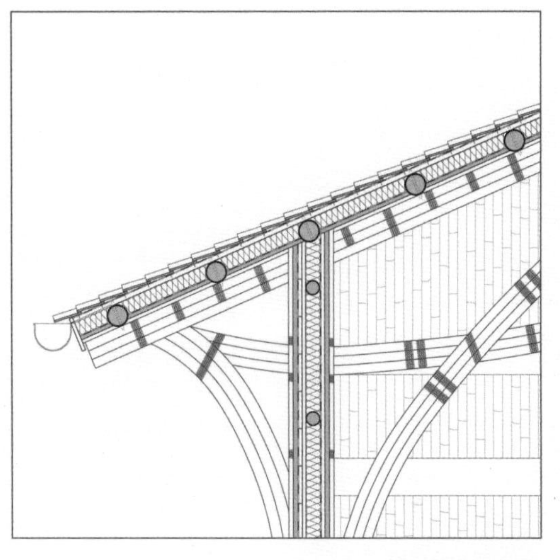

檐口柱节点构造
Construction of eaves pillar nodes

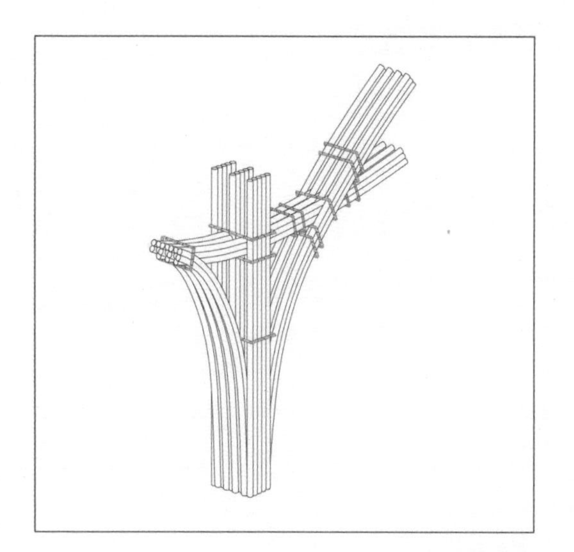

竹构节点
Nodes of bamboo structures

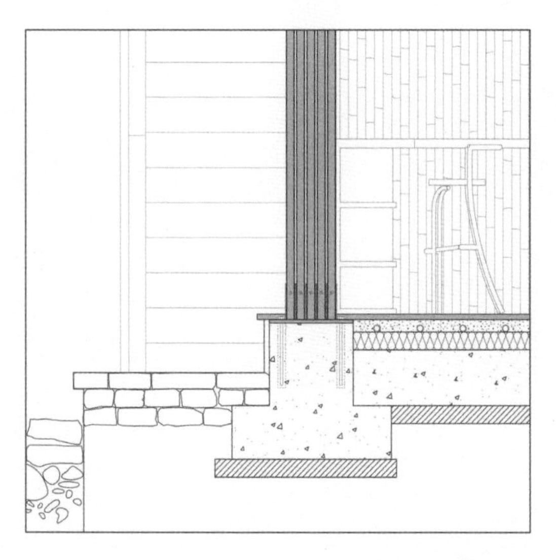

竹－砼基础构造
Construction of bamboo and concrete basement

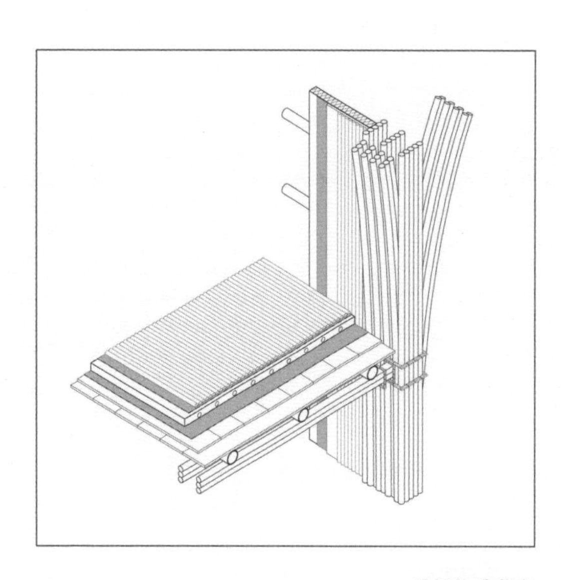

楼板构造节点
Nodes of floor construction

Group 8/IUAV 第八组 威尼斯建筑大学

3.8 重塑土家遗产
Reframing Tujia's Legacy

学生：阿莱格拉·泽恩、阿尔伯特·坎顿
指导老师：阿尔多·艾莫尼诺、恩里科·丰塔纳里、朱塞佩·卡尔达罗拉

Students: Allegra Zen, Alberto Canton
Tutors: Aldo Aymonino, Enrico Fontanari, Giuseppe Caldarola

“重塑土家遗产”项目延续了传统的方法，保留了古盐道历史上土家族传统的建筑元素。通过重新分配空间构形，取得两河口老街中土家建筑公共的外部村落空间和私人的内部家庭空间之间的平衡。整个项目的改造设策略分为三个部分：1. 通过重新平衡更多公共或更私人的内部家庭空间之间的直接或间接关系，以获得更现代的生活条件；2. 通过移除部分传统房屋的主立面，来添加直接属于介于家庭成员和村民公共生活的微尺度小领域空间；3. 通过填充新的建筑材料和使用新的技术手段，为传统民居提供更多的自然采光和增加主动式空调系统。

The "Reframing Tujia Legacy" project follows a conservative approach: all the Tujia traditional architectural elements of the ancient Salt Road in the history are maintained. By redistributing the spatial structure, a balance can be achieved between the public external village space of Tujia buildings and the private internal family space in the Lianghekou old street. The renovation strategy of the entire project is divided into three parts: 1. More contemporary living conditions are obtained through rebalancing the direct or indirect relationship between more public or more private internal domestic spaces; 2. through partially moving the main facade of the traditional house, adds micro-scale space directly belongs to the collective realm of the family members and villagers; 3. through in-filling new architectural disposals and technical means, the design is capable of granting natural air conditioning and solar enlightenment for traditional houses.

● 规划策略
Planning strategies

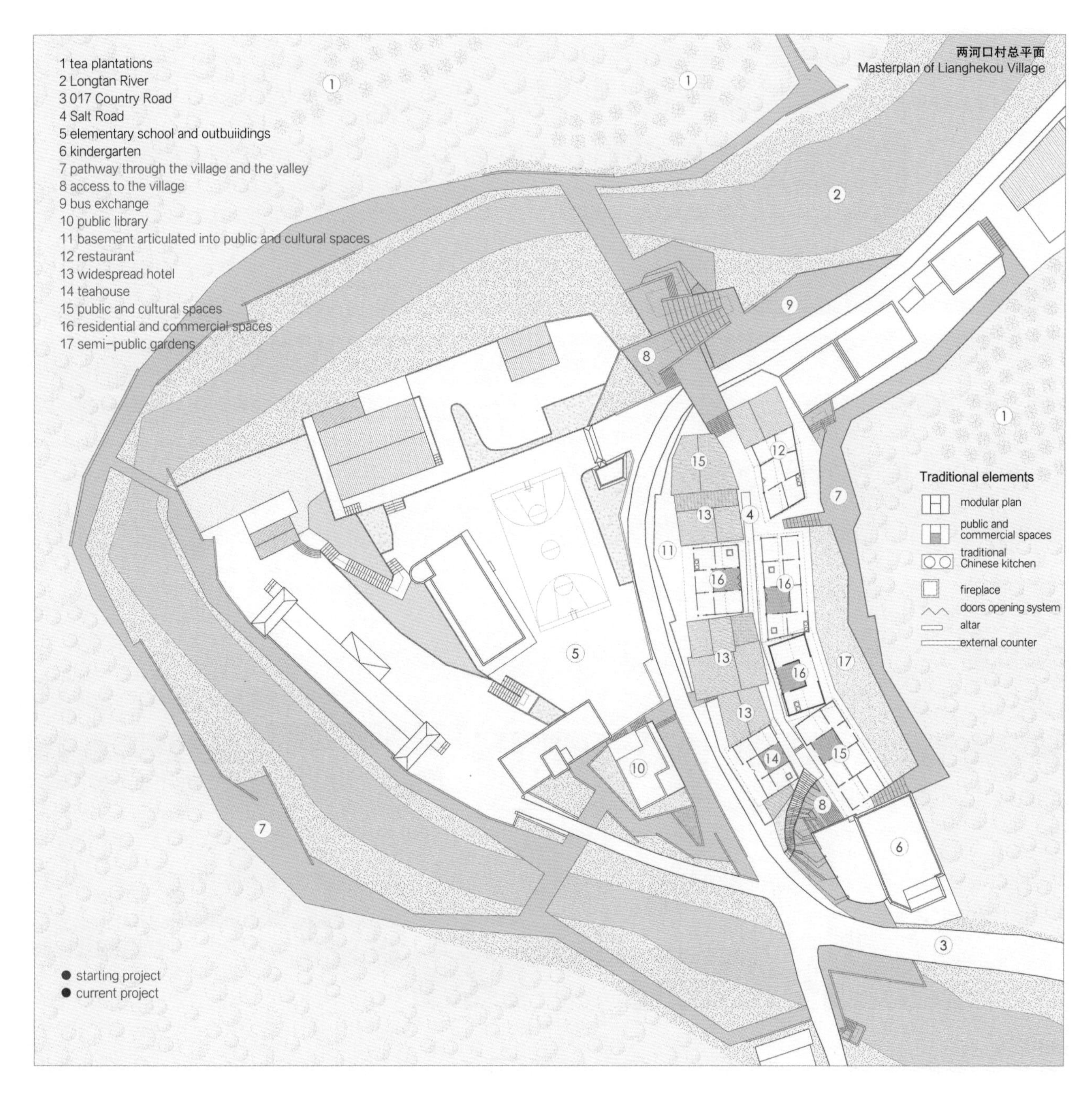

● 建筑设计策略
Architectural design strategies

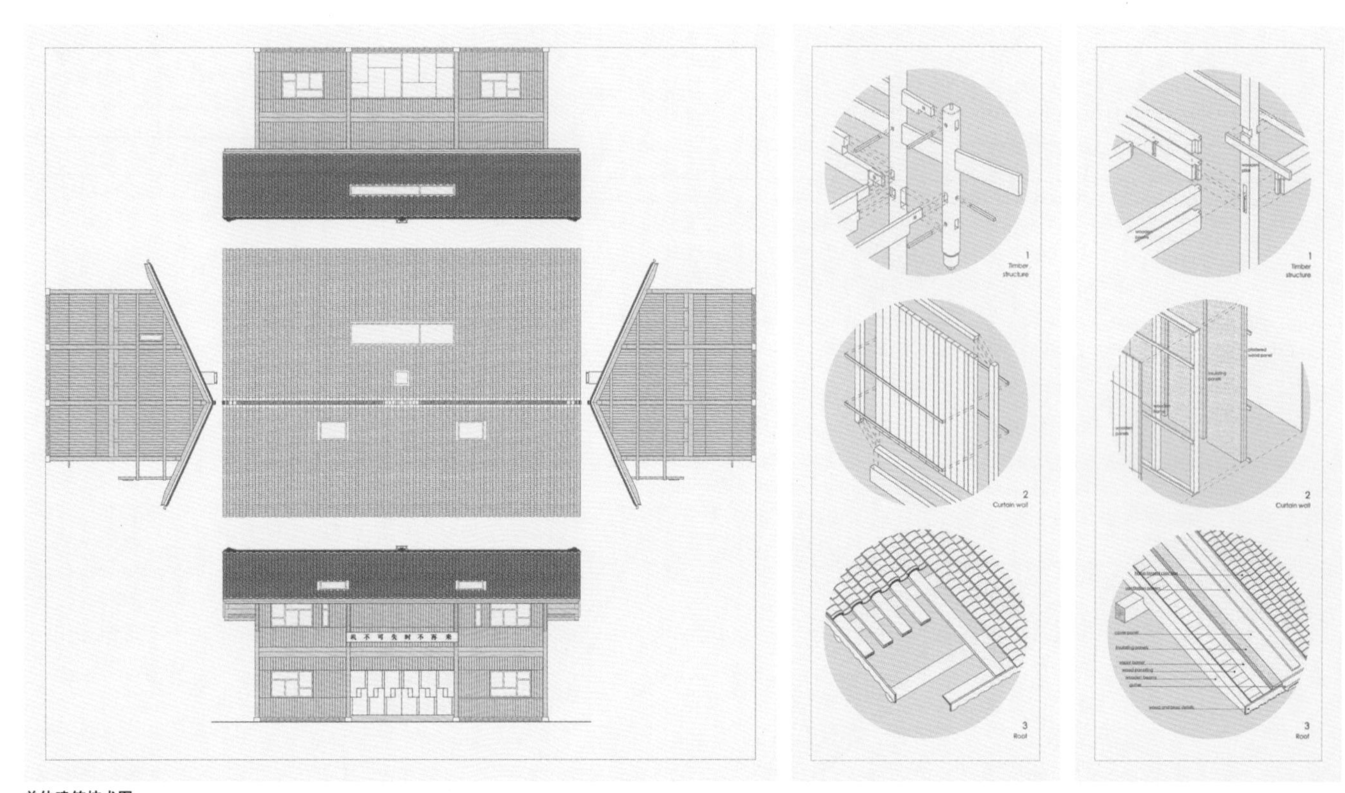

单体建筑技术图
Technical drawings of sigle building

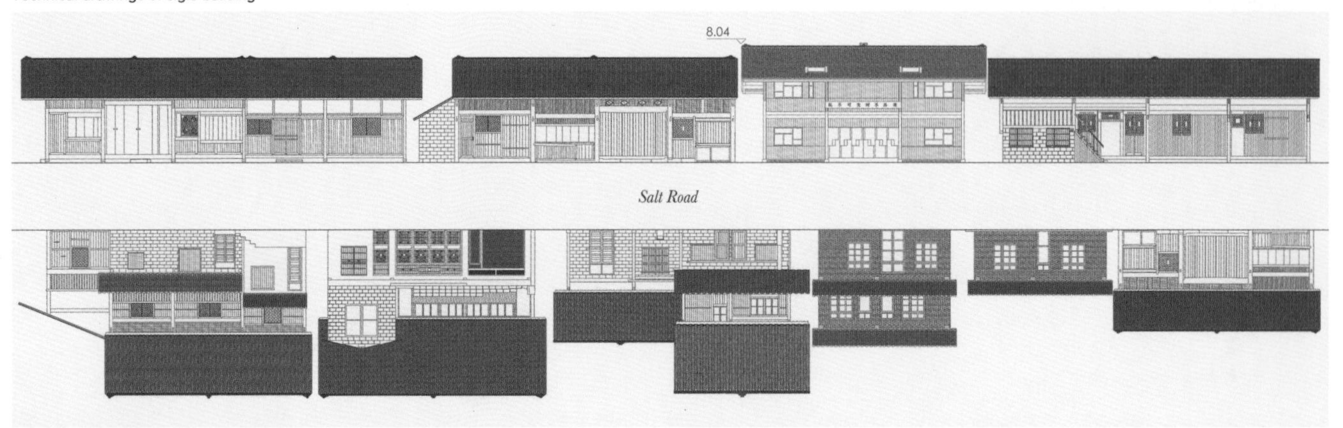

两河口老街街道立面
Street elevation of Lianghekou old street

改造建筑平面图
Plan of the renovated building

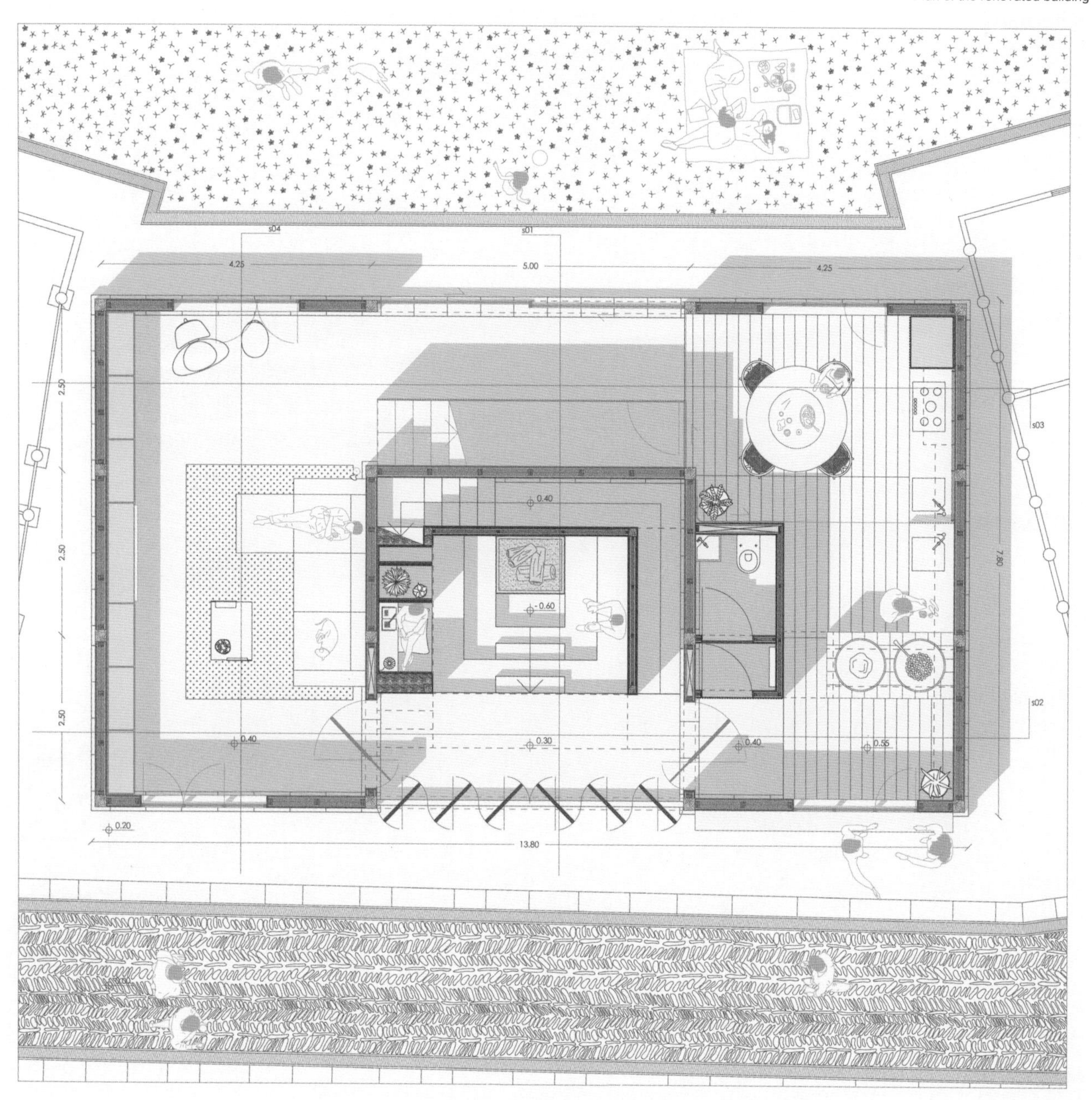

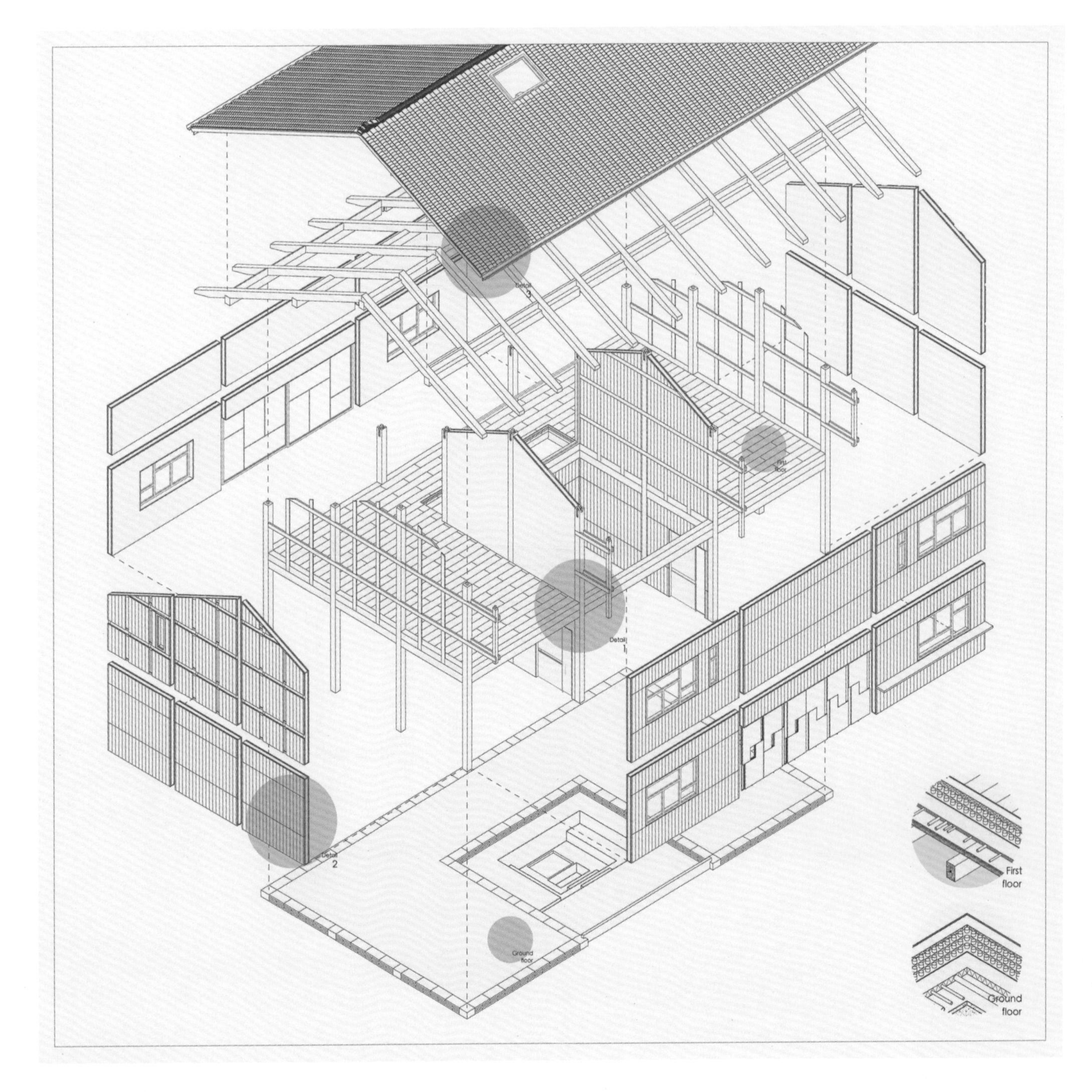

改造建筑结构拆解图
Structure exploded view of the renovated building

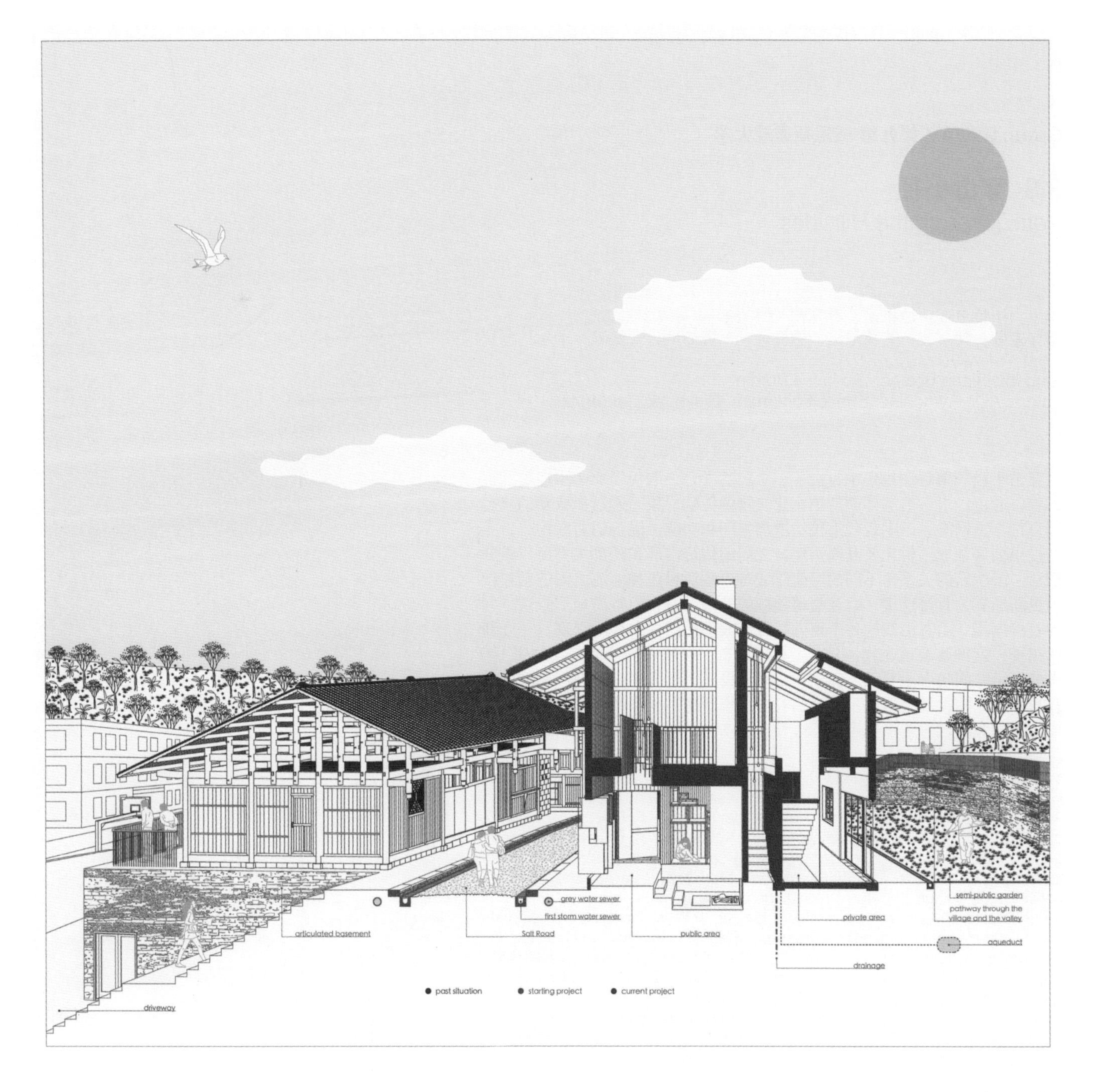

改造建筑剖透视图
Sectional perspective of the renovated building

Group 9/IUAV 第九组 威尼斯建筑大学

3.9 穿山小径
Path Through the Hillsides

学生：伊拉里亚·波提诺、乔凡娜·博尔丁
指导老师：阿尔多·艾莫尼诺、恩里科·丰塔纳里、朱塞佩·卡尔达罗拉

Students: Ilaria Bottino, Giovanna Bordin
Tutors: Aldo Aymonino, Enrico Fontanari, Giuseppe Caldarola

“穿山小径”项目在村庄中打造了一个新的平台，取代了原本不连贯的嵌入物，通过一道新的墙体塑造填充建筑物的传统形象。通过形成新的步行和自行车道网络，产生替代连接，增加空间和功能，混合语言和用户，重新设想公共领域。村庄及其周边地区之间的传统内部和外部关系可以重新激活，产生新的关系。不同的可能路线可以激活村庄和河流之间的新连接或更新原有的直接连接，以及盐道领域的基础设施和山谷的地形景观之间的直接连接。山谷沿线的各个地方都可能成为热点标志和激活剂。一个新的裙楼，就像建筑填充物一样，为村庄创造了一层新的地下室。村庄 100 米的连续空间包含了新的集体功能空间和设施，形成了一堵使用典型的瓦盘技术建立的新墙。新的带楼梯的有盖广场塑造了一个亲密的关系空间，成为别墅和裙楼的入口标志。

The "Path Through the Hillside" project works on the new podium for the village by substituting the actual incoherent insertions and realizing a new wall to establish a traditional image of the architectural in-fill. Re-envisioning the public realm through forming new networks of pedestrian and bicycle pathways, generating alternative connections, adding spaces and functions and mix-languages and users. Traditional relationships, both internal and external, between the village and its sorroundings can be re-activated: new ones can be generated. Different possible itineraries can activate new links or renovate original direct connections both between the village and the river and between the territorial infrastructure of the Salt Road and the terraced landscape of the valley. Various places along the valley can become hotspot indicators and activators. A new podium, as architectural in-fill generates a new basement level for the village. A continuous 100 m space for the village contains new collective functions public spaces and facilities, realizing a new wall which use the typical Wapan technique to establish . A new covered plaza with staircases—an intimate livable relational space—marks the entrance to the village and to the podium.

● 规划策略
Planning strategies

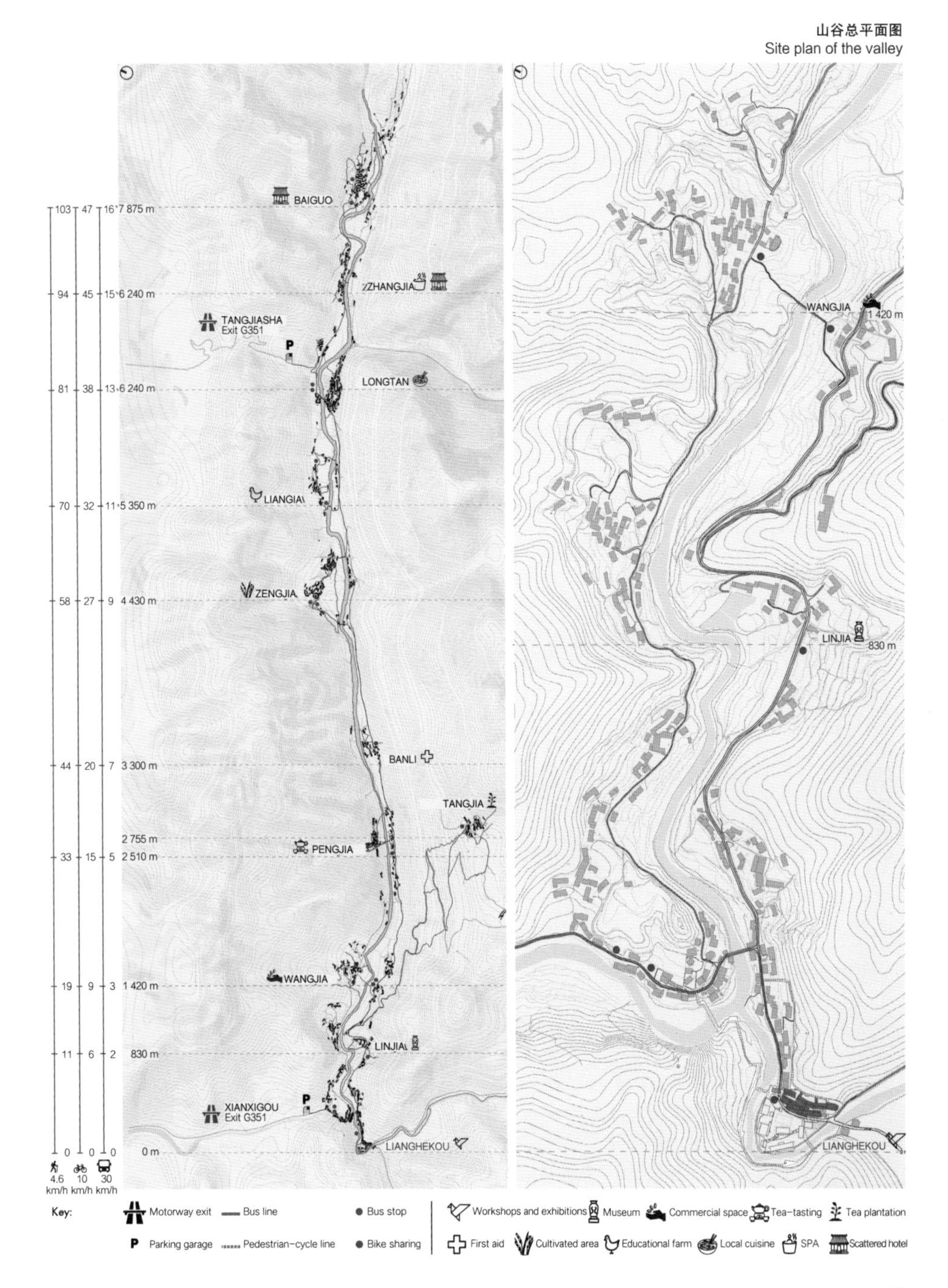

山谷总平面图
Site plan of the valley

穿山路径及平台介入地形的方式
Ways of pathways through the hillside and platforms intervening the terrain

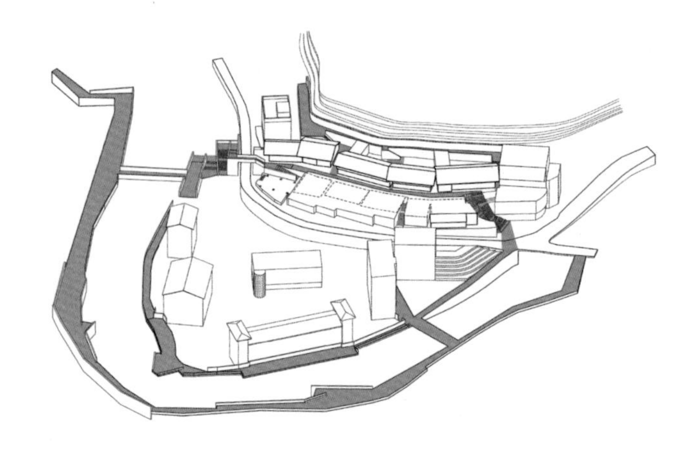

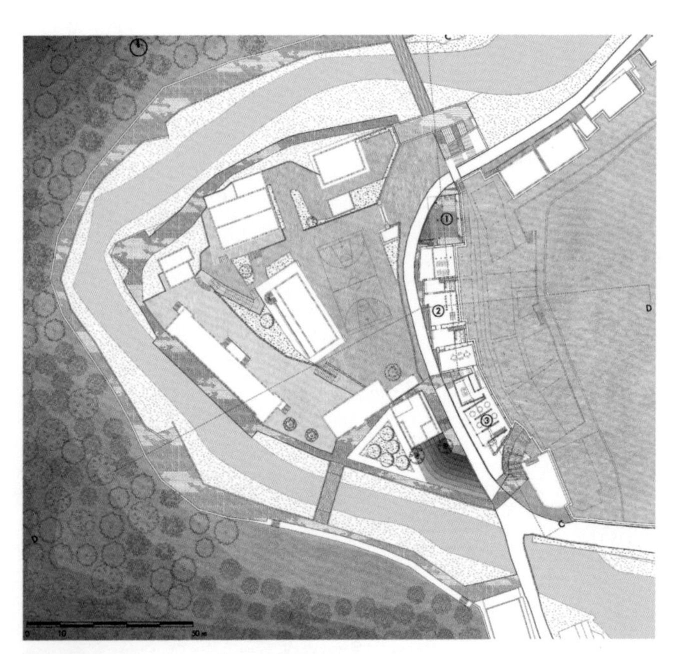

一层平面图
First floor

介入图
Intervention diagrams

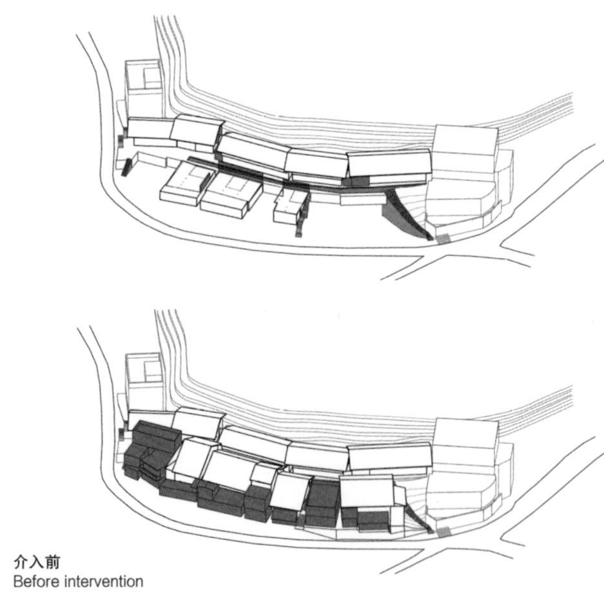

介入前
Before intervention

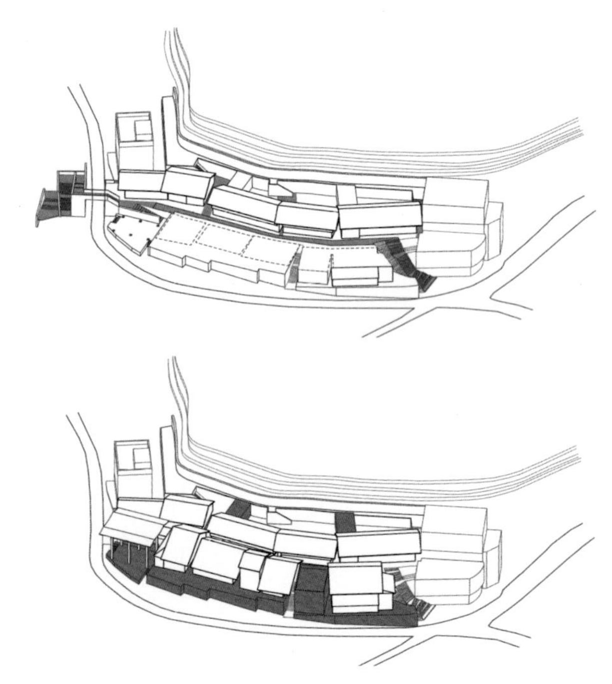

介入后
After intervention

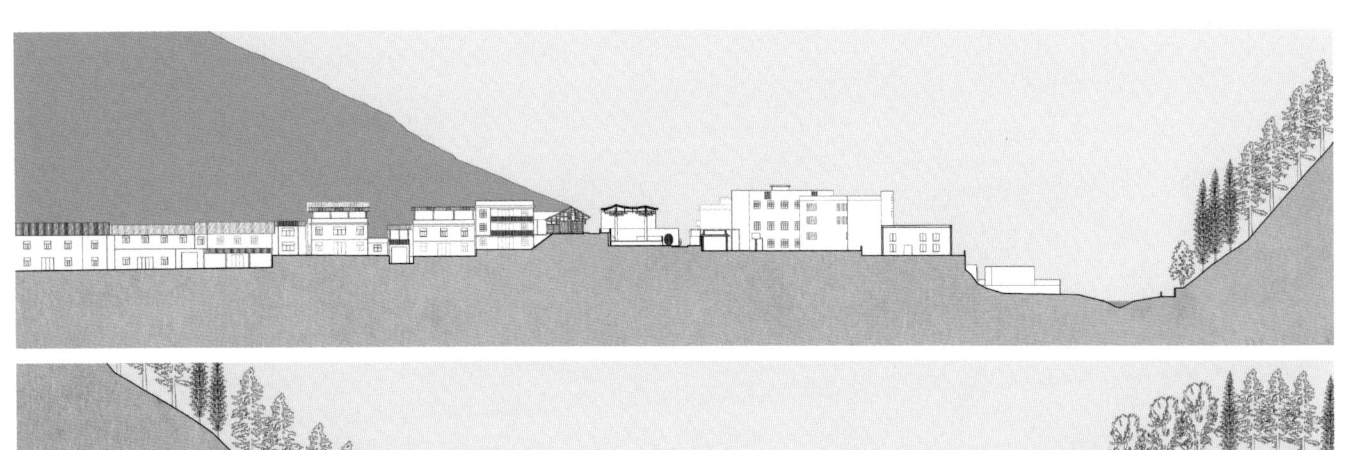

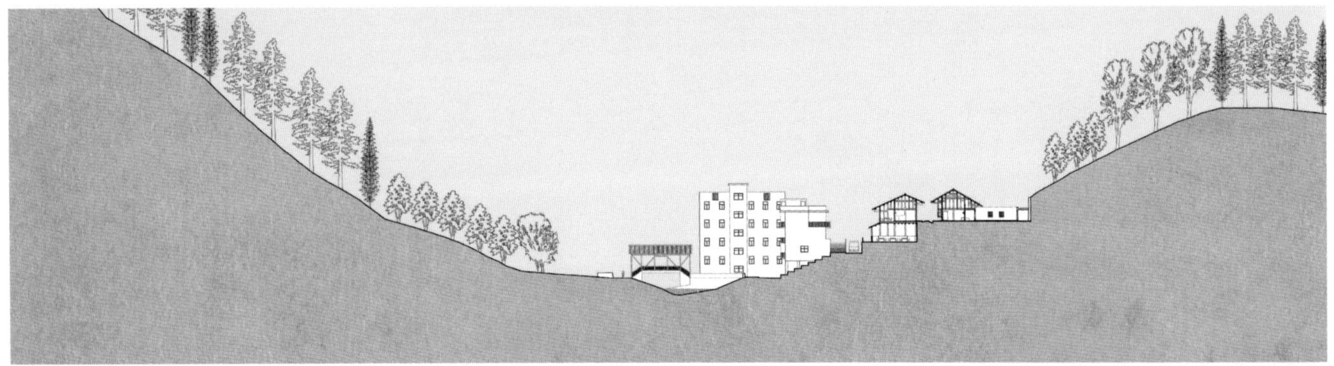

剖面图
Sections

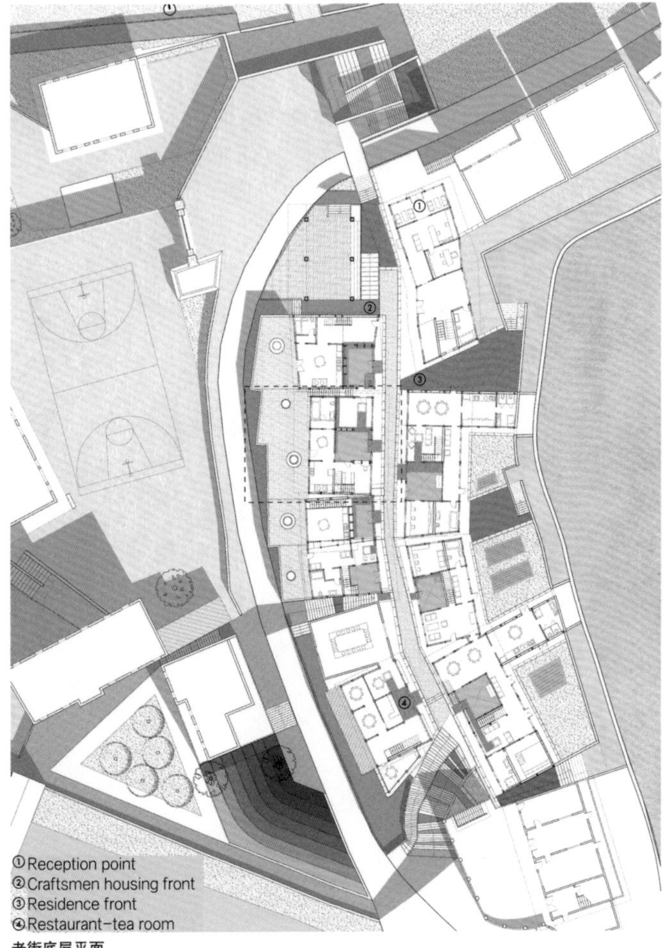

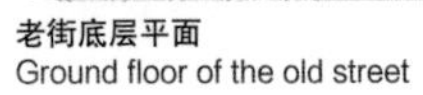

老街底层平面
Ground floor of the old street

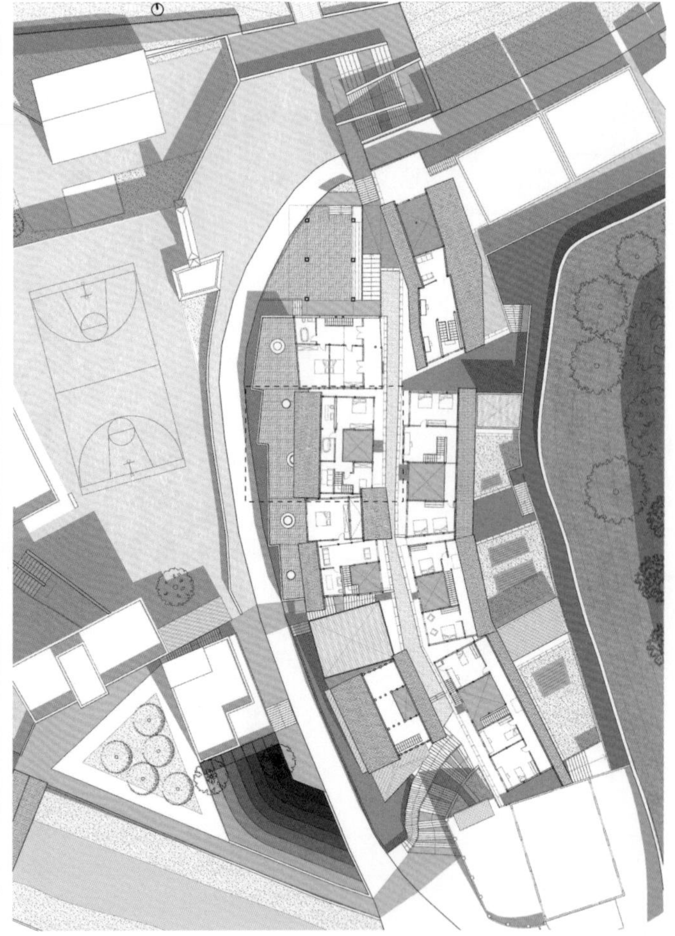

老街一层平面
First floor of the old street

● 建筑设计策略

Architectural design strategies

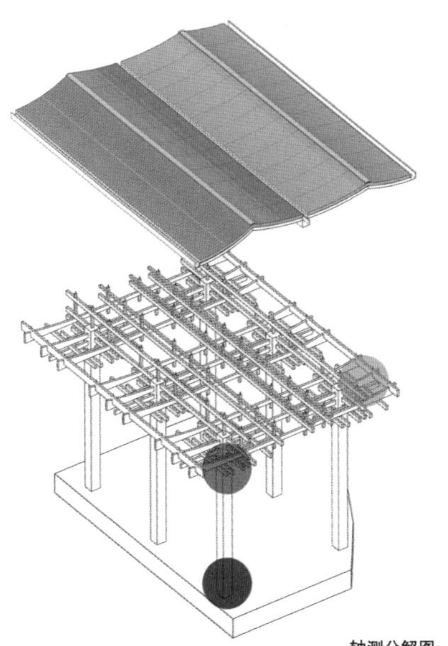

轴测分解图

Axonometric exploded view

覆盖区域俯瞰

Overview of the coverage area

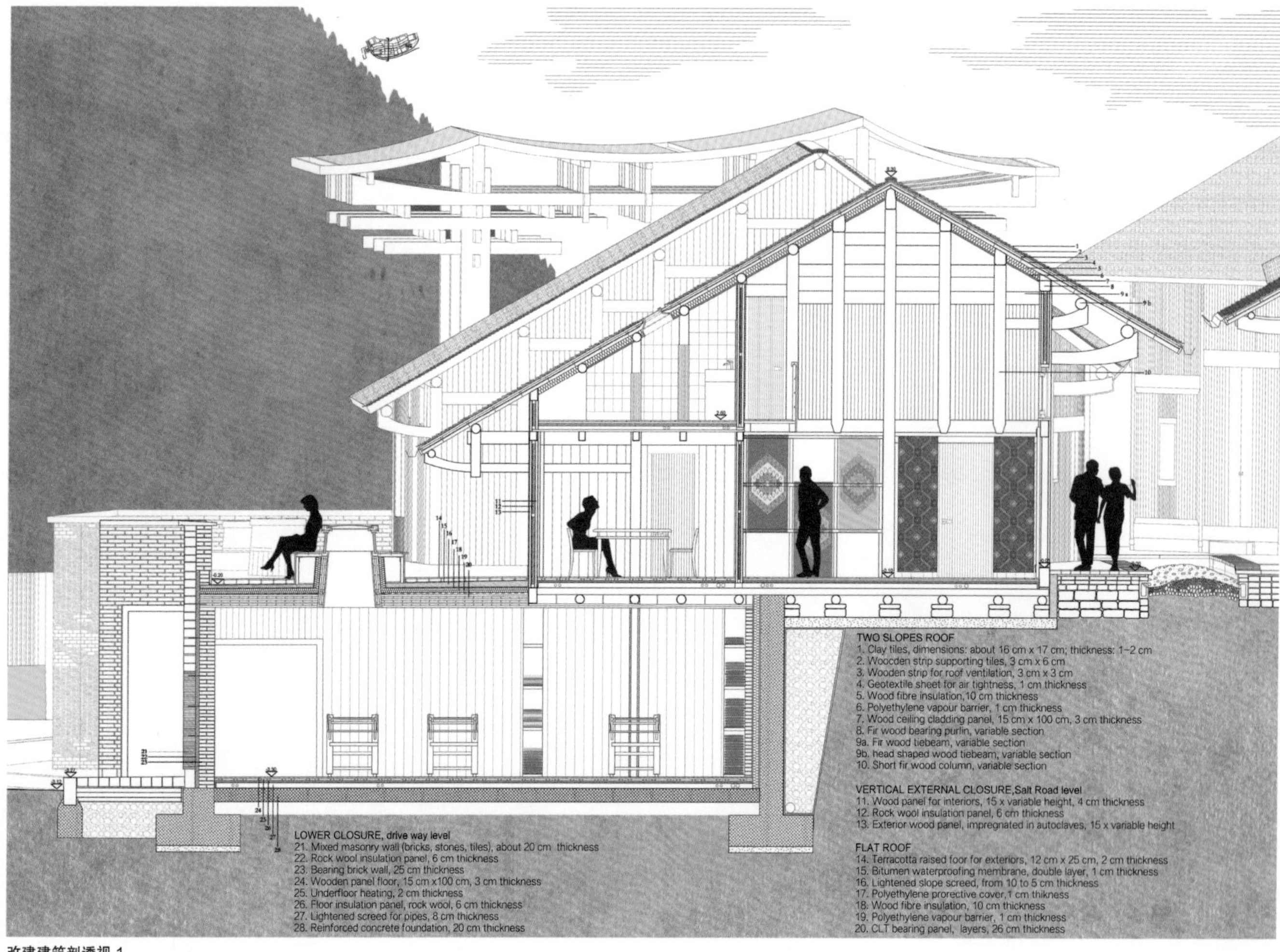

改建建筑剖透视 1

Sectional perspective of the reconstructed building 1

改建建筑剖透视 2
Sectional perspective of the reconstructed building 2

Group 10/IUAV 第十组 威尼斯建筑大学

3.10 土家门户
Tujia Threshold

学生：托马斯·奥尔托兰、贾科莫·雷托雷
指导老师：阿尔多·艾莫尼诺、恩里科·丰塔纳里、朱塞佩·卡尔达罗拉

Students: Thomas Ortolan, Giacomo Rettore
Tutors: Aldo Aymonino, Enrico Fontanari, Giuseppe Caldarola

“土家门户”项目打造了一个山谷九寨中交通枢纽的新的立体交汇点，通过在两河口河谷处打造一个新的片区，重建基础设施，提升景观环境和塑造地标识别。一种区域过滤器调节了内部道路网与2019年刚开通的位于河谷附近的新高速公路之间的关系。一座活力景观塔充当了一个新的换乘点，其两个裙楼上都有停车场。设计引入慢速交通，并通过放弃私家车，选择换乘电动小型汽车和公共汽车穿过山谷，到达所有村庄。景观塔下的一个实验性茶园改造了典型的梯田景观，并在塔、步行区和自行车道网络与村庄之间建立了联系。小公园就在每个村庄的前面。覆盖停车场的传统形状的屋顶，最大限度地减少了新建建筑物对原始环境风貌的影响。

The "Tujia Threshold" project has created a new intersection of transportation hubs among the nine villages in the valley, rebuilding infrastructure, upgrading landscape and shaping landmark through establishing a new area in the Lianghekou valley. A sort of territorial filter modulates the relationship between the internal roads network and the new highway that opened in 2019 and was located near the valley. An energetic tower, with parking lots on the two podium levels, acts as a new interchange point. Slow mobility is introduced and sustained by leaving private vehicles and picking up electric small cars and buses for moving through the valley and reaching all the villages. An experimental tea garden, below the tower, renovates the typical terraced landscape and generates the connections between the tower, the network of pedestrian and bicycle pathways and the villages. Small parking areas are located just in front of each village. Traditionally shaped roofs, covering the parking lots, minimize the impacts of the insertions.

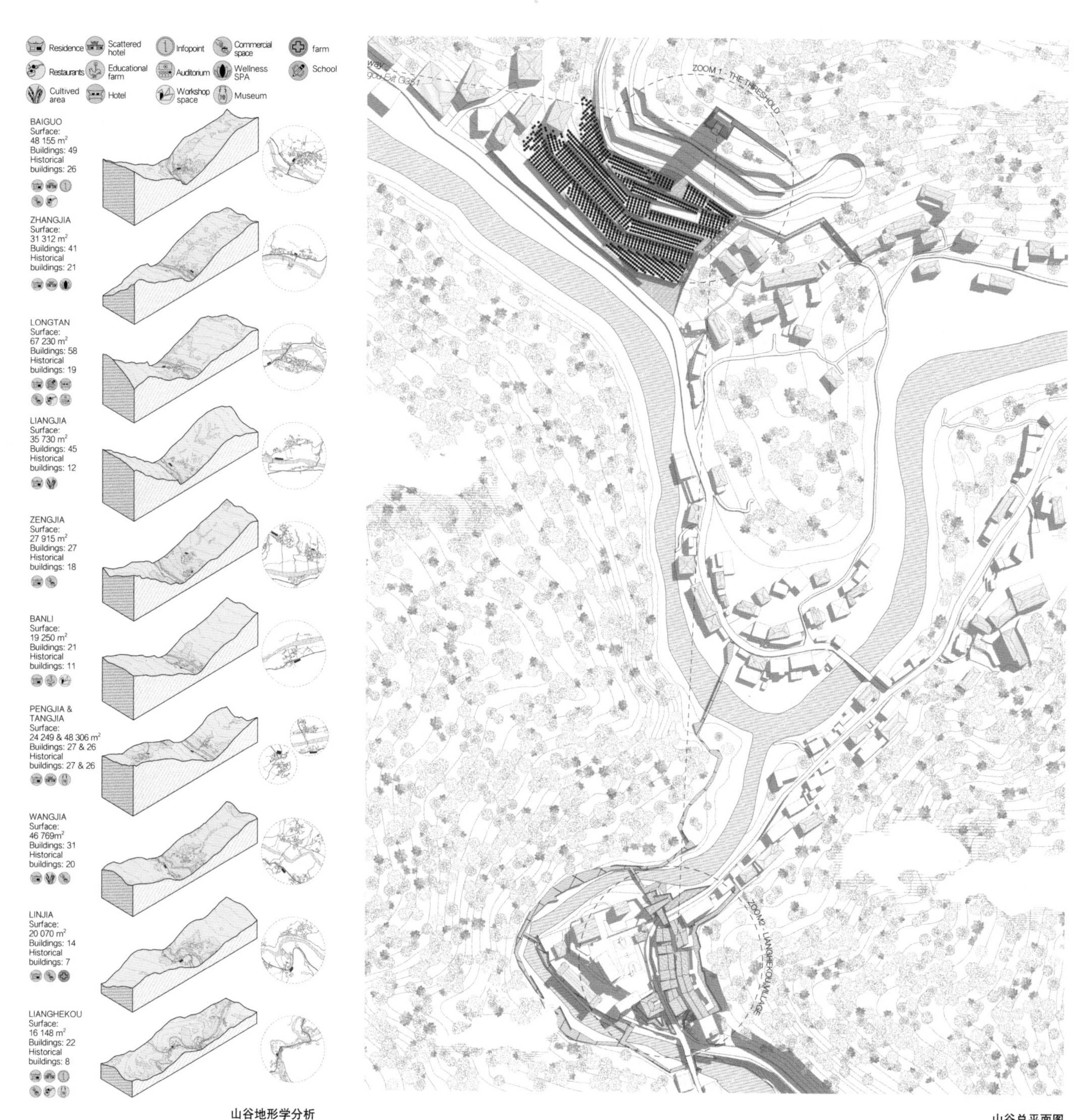

山谷地形学分析
Topographical analysis of the valley

山谷总平面图
Siteplan of the valley

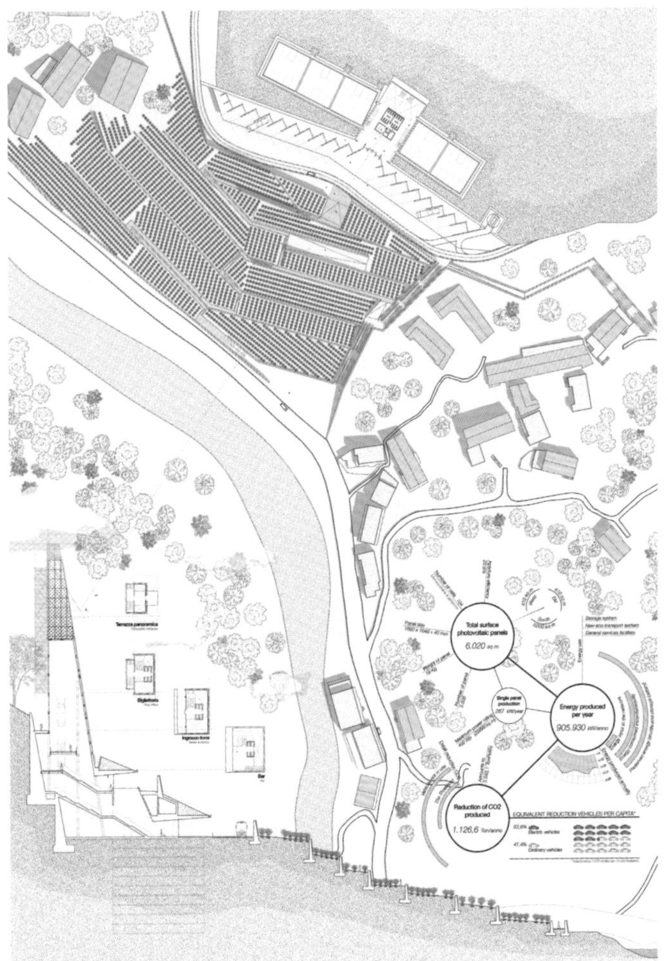

土家门户底层平面及 AA' 剖面
Ground floor plan and AA' section of Tujia Threshold

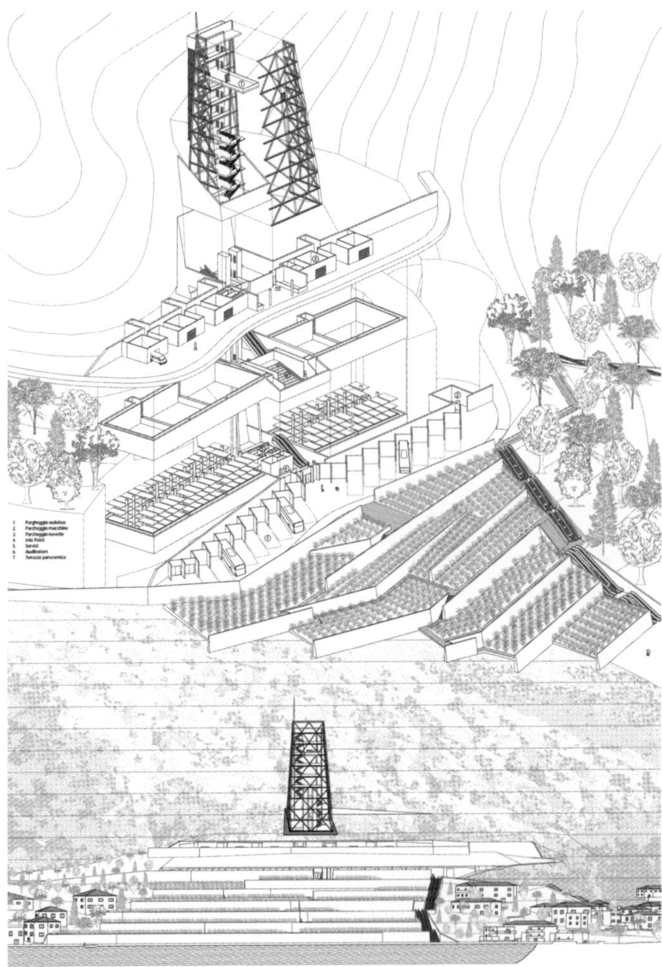

土家门户轴测分解图和立面图
Axonometric exploded view and elevation of Tujia Threshold

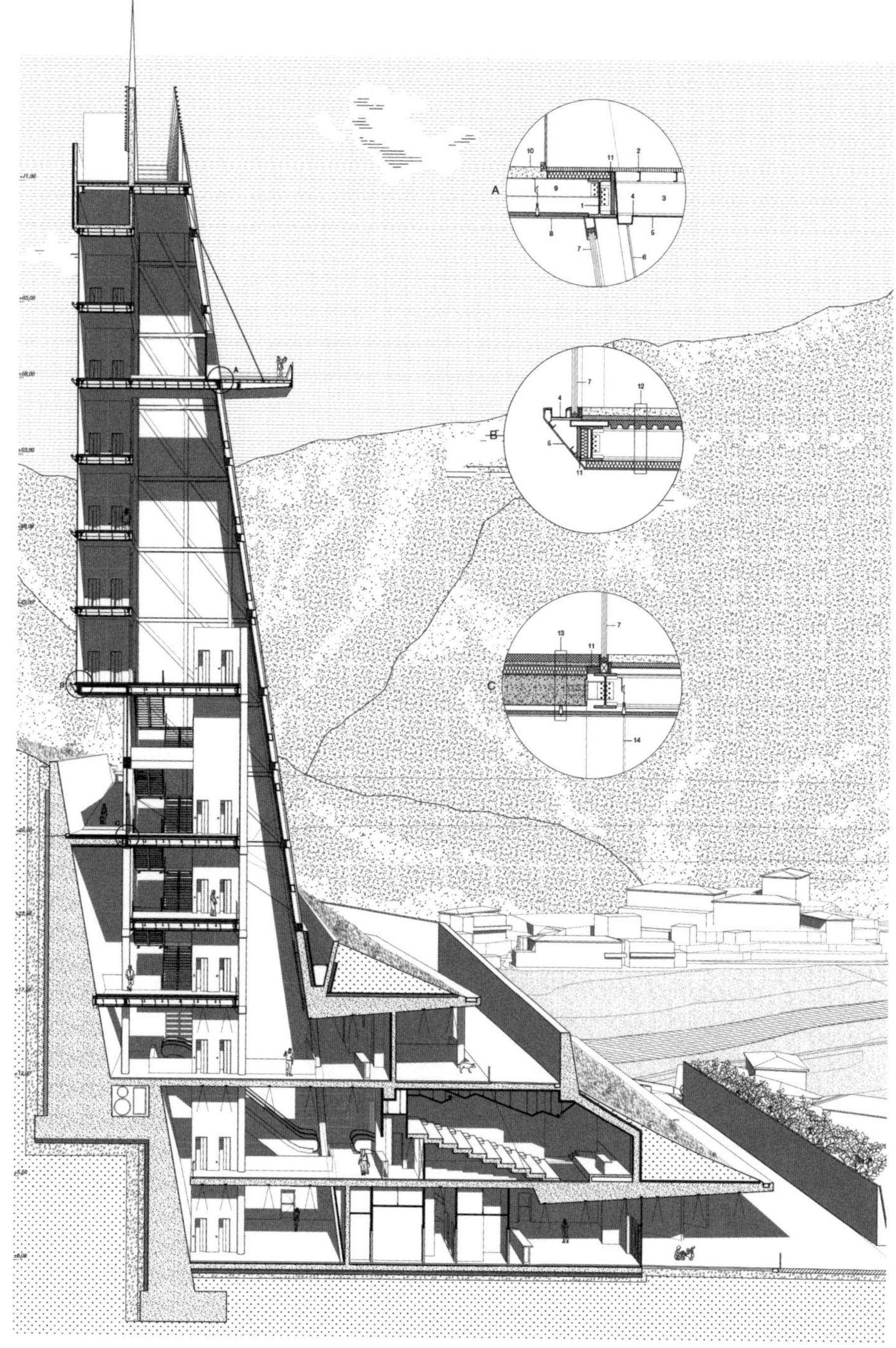

土家门户剖面及结构详图
Section and structural detail of Tujia Threshold

土家门户轴测图
Axonometric drawing of Tujia Threshold

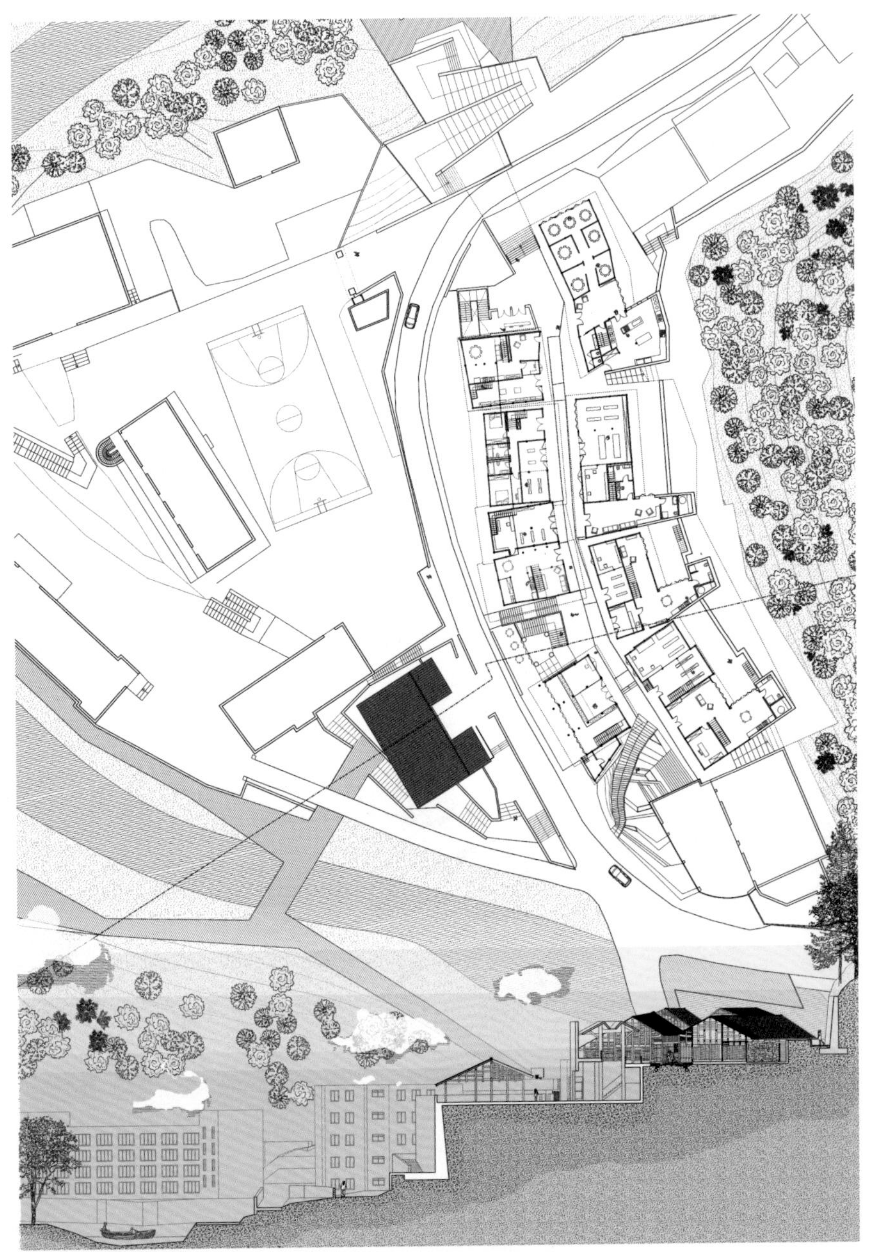

土家门户一层平面及剖面
First floor plan and section of Tujia Threshold

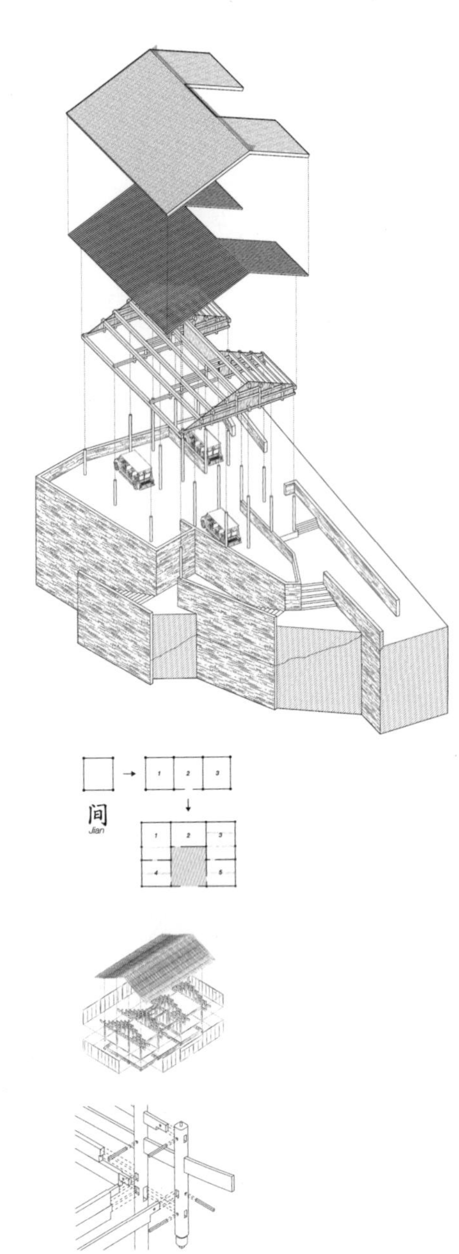

停车场轴测分解图
Axonometric exploded view of car parking

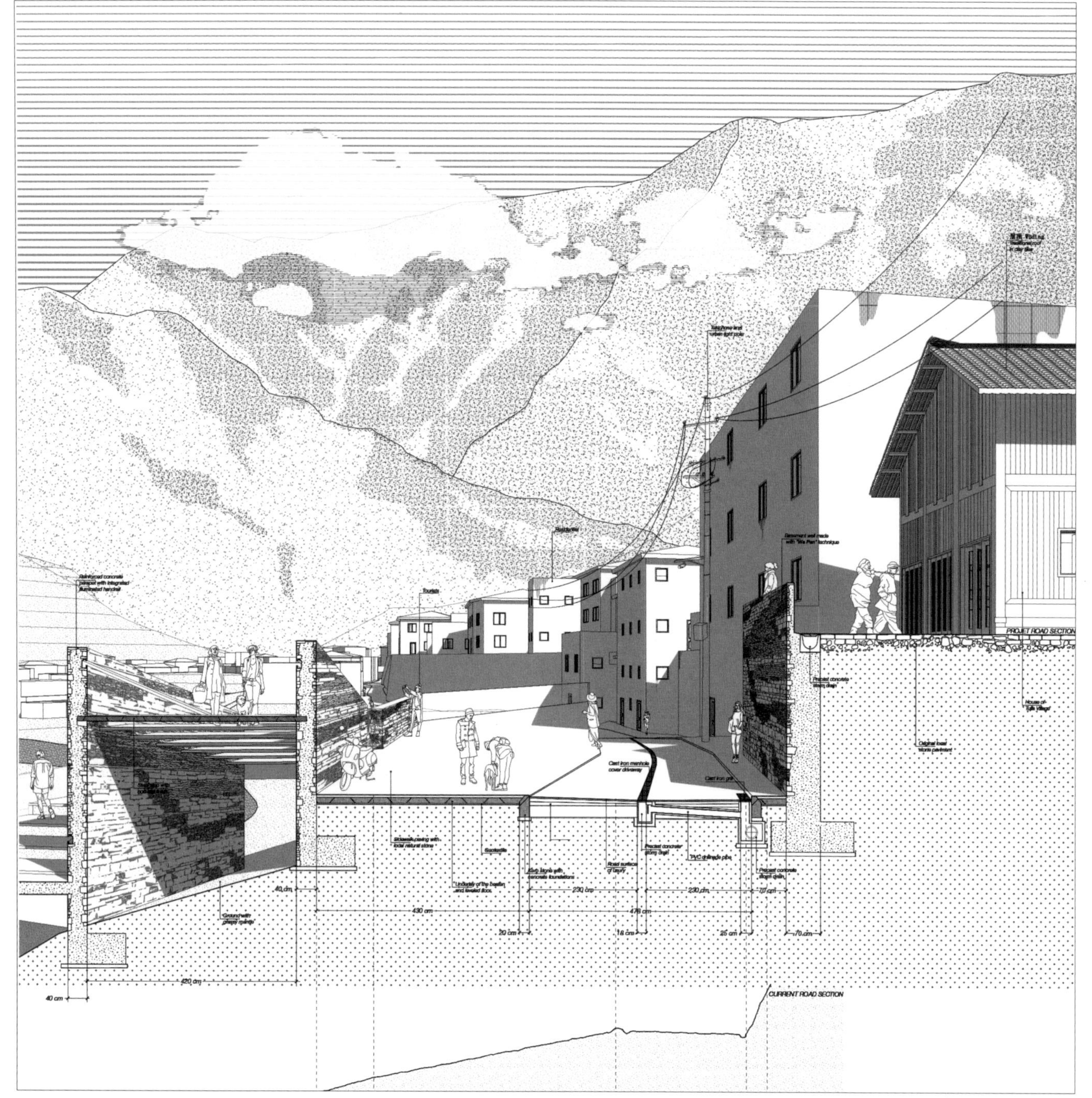

道路剖面：基础设施升级
Road section: upgrading of the infrastructure

4.1 威尼斯建筑大学评图
Review in IUAV

2019 年 6 月 14 日，“两河口：土家盐道古村的再生”国际四校联合研学营阶段性成果汇报与评图在威尼斯建筑大学塔夫里报告厅举行，对过去近两个月的研究设计工作进行汇报、展示与交流。答辩之后，工作营的阶段性成果于 6 月 18 日在意大利雅伦格文化艺术基金会威尼斯禅宫展厅展出，内容包含了两河口的前期调研测绘、主题研究成果以及各校提出的七个精彩设计方案。

On 14th June 2019, "Lianghekou: the Regeneration of an Ancient Tujia Village along Salt Road"—the review of the four universities international joint studio was held in Tufuri Lecture Hall in Università Iuav di Venezia (IUAV), to report, exhibit and communicate the research and design works in the past two months. All of the design panels, including mappings, topics of research as well as the seven projects designed by four universities were exhibited in the Fondazione EMGdotART exhibition hall, Palazzo Zen, on 18th June.

I U A V

Università Iuav di Venezia

LIANGHEKOU

proposte per la rigenerazione di un antico villaggio Tujia lungo la Strada del Sale (Cina)

seminario di presentazione
degli esiti del workshop progettuale
intervengono **Aldo Aymonino**, **Enrico Fontanari**, **Giuseppe Caldarola** Iuav, **Zhang Tong** e **Chuan Wang** SEU, **Baofeng Li** e **Shikuang Tang** HUST, **Dongzhu Chu** e **Chong Gong** CQU

17.6.2019
Palazzo Zen
(EMGdotART Foundation)

Inaugurazione mostra ore 17

威尼斯建筑大学宣传海报 Promotional poster in IUAV
图片来源：威尼斯建筑大学 IUAV 提供

中期汇报答辩前，四校师生在威尼斯商议教学阶段成果的展示方式和后续工作进程。在中西方不同文化遗产背景下对两河口乡村活化的教学再度展开思辨。阶段性成果的汇报是对之前两河口村小组工作进行共同汇总，以各个学校为单位分组进行不同思路的方案深化探索，并在威尼斯建筑大学久负盛名的塔夫里报告厅进行公开汇报答辩。

Before the mid-term review, teachers and students from the four universities discussed the presentation method and subsequent work schedule of the teaching stage results in Venice. The teaching of the revitalization of Lianghekou rural areas in the different cultural heritage contexts in the East and West has once again been discussed. The report on the phased achievements is a joint summary of the previous work of the Lianghekou Village groups, and each university as a group conducts in-depth exploration of different ideas and solutions. The public presentations are conducted in the prestigious Tufuri Lecture Hall in IUAV.

师生讨论在威尼斯教学工作安排

Teachers and students discussed on the teaching schedule in Venice

各组师生在汇报前，集中在威尼斯建筑大学工作学习，进行紧张的方案策划、资料查阅、图纸表达和模型制作等环节的工作。

Before the presentation, the teachers and students of each group worked and studied at Università Iuav di Venezia. This includes tintensive program planning, data search, drawing expression, and model making and other sections.

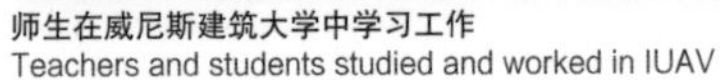

师生在威尼斯建筑大学中学习工作
Teachers and students studied and worked in IUAV

每个设计组进行 20 分钟的阶段成果工作展示，现场各校老师对方案设计进行点评。评图结束后，师生们进行了圆桌对话，共同探讨此次研学营的心得体会，并对后续设计深化、成果的落地和展现做出展望。

Each group gave a 20-minute presentation of stage achievement work and teachers from different schools commented on their scheme design proposals. After the review, the teachers and students held a round table dialogue to discuss the experience of this joint studio, and made a prospect for the follow-up design deepening, implementation and presentation of the outcomes.

学生在塔夫里报告厅进行方案汇报

Students are giving presentations of their proposals in Tufuri Lecture Hall

● 教师点评
Teachers´ review

在教学的阶段性汇报答辩中，中意师生展开了热烈讨论：设计如何发挥自身潜在的力量来改进和平衡文化遗产与现代旅游业以及城镇化发展之间的矛盾。这也是本次课程教学的出发点和值得持续思考的议题。教学过程中，来自不同文化背景的几所高校师生相互交流，思想碰撞，从场域生态、遗产保护、功能业态、活化改造、村镇公共空间营造等多个层面，对两河口村未来的发展提出了多种模式和策略的探索。在威尼斯禅宫的展览不是最终成果的展示，而是对这一富有价值探索的有力推进，希望通过展览与实践，使得教学中的思考能够产生更为深远而广泛的影响。

教师评图环节
Teachers' review session

师生在威尼斯建筑大学合影 Group photo of teachers and students at IUAV
图片来源：威尼斯建筑大学（IUAV）提供

During the phased presentation and defense of teaching, Chinese and Italian teachers and students engaged in lively discussions: how to exert potential architectural power to improve and balance the contradiction between cultural heritage and modern tourism under urbanization development. This is also the starting point of this course teaching and a topic worth continuous consideration. In the teaching process, teachers and students from different cultural backgrounds communicated with each other, and ideas were exchanged. From the perspectives of field ecology, heritage protection, functional formats, activation and reconstruction, and public space construction of villages and towns, various models and strategies were put forward for the future development of Langhekou Village. The exhibition in Palazzo Zen was not the display of the final result, but a powerful promotion of this valuable exploration, hoping that through the exhibition and practice, the continuous thinking of education can have a more profound and extensive impact.

4.2 禅宫课程展览
Workshop Exhibition in Palazzo Zen

答辩之后，工作营的阶段性成果在意大利雅伦格文化艺术基金会威尼斯禅宫展厅举办了展览，内容包含了两河口的前期调研测绘、六个主题研究成果以及来自四所院校的各组提出的七个主题设计方案。雅伦格文化艺术基金会主席马里诺·福林教授、威尼斯建筑大学副校长恩里科·丰塔纳里教授和建筑系主任阿尔多·艾莫尼诺教授、东南大学建筑学院院长张彤教授 以及四所高校的师生近百人齐聚雅伦格文化艺术基金会禅宫展厅，开启为期十天的成果展览。

After the review, the preliminary results of the joint studio were exhibited in the Palazzo Zen in Venice funded by the Fondazione EMGdotART, including the survey measurements of Lianghekou Village, six thematic research results, and seven different design proposals made by each group from the four universities. Prof. Marino Folin, Chairman of the Fondazione EMGdotART, Prof. Enrico Fontanari, Vice Rector of the IUAV, Prof. Aldo Aymonino, Head of the Department of Architecture(IUAV), Prof. Zhang Tong, Dean of the School of Architecture(SEU), as well as nearly 100 visitors, gathered in the Palazzo Zen funded by the Fondazione EMGdotART. This exhibition lasted for about 10 days .

中期评图在禅宫展览现场
Mid-term evaluation chart in Palazzo Zen exhibition site

● 课程展览活动现场
Course exhibition activity site

参展师生在禅宫展览现场工作合影
Group photo of participating teachers and students working at the exhibition site in Palazzo Zen

TRATTO DA "LA SYLPHIDE"
DI MARIA TAGLIONI DEL 1832
SULL'IMPOSSIBILE AMORE
TRA UN ESSERE UMANO
E UNA CREATURA
SOPRANNATURALE

Martedì 18 Giugno 2019
www.gazzettino.it

PALAZZO ZEN Nella foto la presentazione dell'iniziativa

Progetti a confronto su un villaggio cinese

RIGENERAZIONE

VENEZIA "Lianghkou. Rigenerazione di un antico villaggio Tujia lungo la Strada del Sale" (Cina) è il titolo della mostra inaugurata ieri a Palazzo Zen, nello spazio espositivo della Fondazione EMGdotART. «Una mostra importante per la qualità del lavoro svolto» ha sottolineato Marino Folin, nel presentarla esprimendo un parere molto positivo sui risultati. Quattro università hanno infatti partecipato e i loro progetti sono stati elaborati in forma congiunta da studenti e docenti dell'università Iuav e tre università cinesi: Southeast University of Nanjing (Seu – Nanchino) / Huazhong University of Science and Technology (Hust – Wuhan) / Chongqing University (cqu – Chongping City). I lavori sono stati coordinati dai docenti Aldo Aymonino, Enrico Fontanari, Giuseppe Caldarola (Iuav), Zhang Tong e Chuan Wang (Seu) , Baofeng Li e Shikuang Tang (Hust), Dongzhu Chu e Chong Gong (Cqu). «Fondamentale - ha concluso Folin - è stato mettere a confronto le due culture e i diversi approcci. Uno scambio proficuo per soluzioni più giuste, non solo per recuperare e ammodernare, ma per mantenerne lo spirito e la vita sociale».

Lianghkou è un villaggio commerciale lungo l'antica strada del sale che si è formato durante le dinastie Ming e Qing. Costruito seguendo un modello morfologico lineare dei villaggi commerciali di cui conserva l'immagine tradizionale, il villaggio rappresenta un tassello di un sistema insediativo composto da piccoli villaggi. Lungo il tempo vi è stato un decremento della popolazione ed un avanzato stato di degrado e abbandono. Attualmente vi è un'attenzione sul valore patrimoniale di questi luoghi e si sperimentano nuove possibilità di turismo.

I villaggi Tujia rientrano nell'individuazione dei possibili attrattori turistici, grazie al loro inserimento negli elenchi dei luoghi patrimoniali nazionali e alla costruzione di una nuova infrastruttura autostradale. Dai primi di maggio studenti e docenti delle quattro università hanno proseguito le attività di studio e progettazione con diverse proposte di intervento e rigenerazione di questo sistema insediativo. Gli esiti progettuali finali saranno esposti nel 2020 in una mostra a cura della Fondazione EMGdoART, a Palazzo Zen, in concomitanza con la prossima Biennale di Architettura.

Maria Treresa Secondi

Carlo & Giorgio presto al Lido con la loro "Famiglia Baldan"

威尼斯当地媒体《Il Gazzettino 》日报于 6 月 18 日对禅宫的展览进行了现场专题报道
Newspaper Il Gazzettino, a local media in Venice, took a live report on the exhibition on 18th June
图片来源：威尼斯当地媒体 Il Gazzettino 日报 6 月 18 日刊

第五章 第十七届国际威尼斯建筑双年展
CHAPTER 5 17TH LA BIENNALE DI VENEZIA

5.1 官方平行展——两河口，一个土家会聚之地的再生
Collateral Event—Lianghekou, a Tujia Village of Re-Living-Together

● 展览信息
Exhibition information

展览时间 :2021 年 5 月 22 日—11 月 21 日
Time: 22nd MAY to 21st Nov, 2021

展览地点：意大利威尼斯卡纳雷吉欧区 4924 号禅宫
Site: Palazzo Zen, Cannaregio 4924, Venice, Italy

策展人 Cruators:

张彤
Zhang Tong

恩里科 · 丰塔纳里
Enrico Fontanari

阿尔多 · 艾莫尼诺
Aldo Aymonino

参展单位 Organizers:

东南大学
SEU

威尼斯建筑大学
IUAV

华中科技大学
HUST

重庆大学
CQU

宣恩县
Xuan'en County

雅伦格文化艺术基金会
Fondazione EMGdotART

威尼斯卡纳雷吉欧区4924号禅宫展览海报
Cannaregio 4924, Venice, Palazzo Zen exhibition poster
图片来源：http://www.emgdotart.org

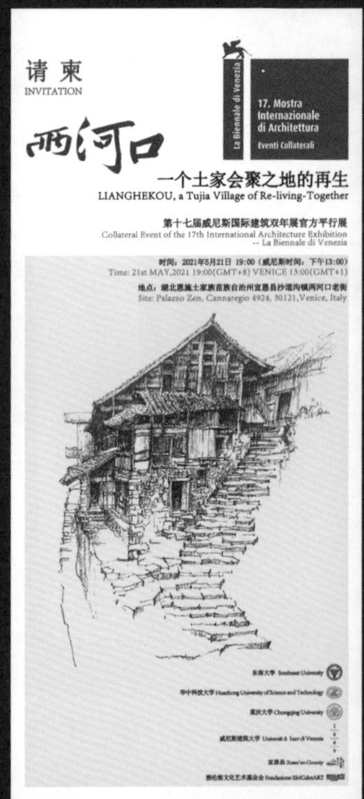

中国湖北宣恩两河口村活动请柬
Lianghekou Village, Xuan'en, Hubei, China, event invita
图片来源：东南大学中华民族视觉形象基地提供

两河口老街和威尼斯开幕式现场互动连线
Live broadcast of the opening ceremony between Lianghekou old street and Venice
图片来源：亚伦格文化艺术基金会提供

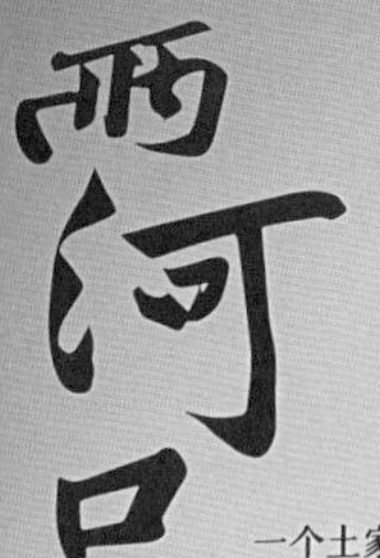

一个土家会聚之地的再生

Promoter

Curators

Participants

Sponsors and Supporters

LIANGHEKOU: a Tujia Village of Re-Living-Together

Collateral Event of the 17 th International Architecture Exhibition – La Biennale di Venezia

Time
22nd May - 21st Nov. (Close on Tue)

Venue
Palazzo Zen
Cannaregio 4924, 30121 Venezia

La Biennale di Venezia
17. Mostra Internazionale di Architettura
Eventi Collaterali

Lianghekou, located in Xuan'en County, Enshi Tujia and Miao Autonomous Prefecture, China, was a commercial village along the ancient Salt Road which was a significant economic link in the Ming and Qing dynasties. The well-preserved one-hundred-meter street at the core of this village typically embodies the tectonic characteristics of Gan-lan wooden construction system, especially Tujia tradition of stilt house along this historic path. As a result of rapid urbanization and population emigration, the once thriving village, among many in the region of Wuling Mountains, is suffering from the economic decline and social exclusion.

At present, the once hidden Salt Road valley, where Lianghekou is located, is embracing an opportunity of attracting the attention from the out world on account of the rapid development of the national traffic network nearby and the drastic increase of domestic tourism in China. How to regenerate this declined village along the ancient Salt Road and convert it to be a vigorous place for people to RE-LIVING-TOGETHER? How to address an architectural resilience that supports the economic advantages of tourism development, while celebrating the authentic values and cultural identities of its historical legacies? This exhibition is not only concerning the exciting, unique project itself, but also aims to explore a feasible solution to the common challenge faced by human settlements with cultural heritages worldwide.

展览开幕式两河口老街活动现场
The scenario of Lianghekou old street during the exhibition opening ceremony

The official collateral event of the 17th International Architecture Exhibition—La Biennale di Venezia "Lianghekou, a Tujia Village of Re-Living-Together" co-curated by Professor Zhang Tong, School of Architecture(SEU), Professor Aldo Aymonino and Professor Enrico Fontanari (IUAV), opened on 21st May 2021. The exhibition was held at the Palazzo Zen in Venice, and one ceremonial gala was launched at the same time in the Lianghekou Village, Xuan'en County, Enshi Tujia and Miao Autonomous Prefecture, Hubei Province. A number of media platforms, such as CNR International Online, Hubei Satellite TV, China Daily, China News Network, Changjiang Cloud, Changjiang Daily, Xuan'en Cloud, Facebook, Youtube, Bilibili, etc., have been bilingually spread to the world.

时下，乡村旅游的发展为两河口提供了振兴的契机，但同时也带来了新的挑战：传统村落如何在迅猛的旅游业发展中，避免浅表性的消费，保护地方遗产的原真性与独特性，使得沉睡的村庄得以再生。这不仅仅是我们在两河口遇到的问题，也是包括威尼斯在内的人类文化遗产共同面临的挑战。

The recent development of rural tourism provides Lianghekou a chance for revitalization. Then it also brings the challenge: how to avoid superficial tourist consumption and conserve the authentic values and identities of the local legacies in the rapid development of tourism industry, so as to regenerate the declined village? That is not only a question here at Lianghekou, but also the challenge faced by all human heritages including Venice.

会聚・土家长桌宴
Gathering・Tujia Long-table banquet
图片来源：东南大学中华民族视觉形象基地提供

土家族木结构展厅
Tujia wooden structure exhibition hall
图片来源：雅伦格文化艺术基金会提供

● 威尼斯展厅现场
On site at the Venice exhibition hall

基于国际四校联合研学营的实验性研究和反思性实践，以“两河口，一个土家会聚之地的再生”为主题，登陆2021年第十七届威尼斯建筑双年展，成为15个官方平行展之一。2021年5月22日到11月21日，在意大利威尼斯卡纳雷吉欧区禅宫进行为期半年的展览，向国际社会展示一个具有持久价值的乡村振兴和文化复兴的中国案例。

土家族自然生境的展厅
Natural habitat exhibition hall of Tujia nationality

Based on the experimental research and reflective practice of the four universities international joint studio, with the theme of "Lianghekou, a Tujia Village of Re-Living-Together" was held at the 17th International Architecture Exhibition—La Biennale di Venezia in 2021, it became one of the 15 official collateral events. From May 22 to November 21, 2021, this half-year exhibition was held at the Palazzo Zen in Cannaregio, Venice, Italy, showing the international community a Chinese case of rural revitalization and cultural revival with lasting value.

土家族山谷中九寨格局
The pattern of nine villages in the valley of Tujia nationality

土家火塘生活的展厅
Tujia fire-pits life exhibition hall

展厅入口
The entrance of the exhibition hall

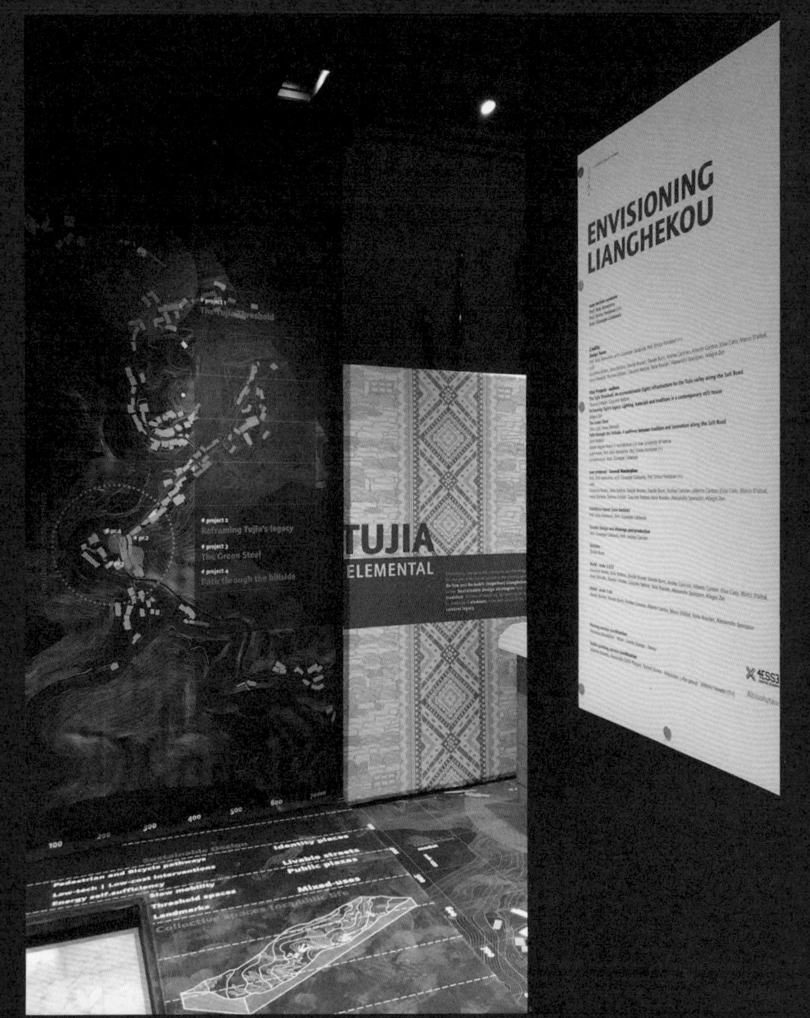

展览所展示的两河口的再生实践是沿着古盐道的整个土家河谷总体活化利用的重要部分。在即将到来的整体土家河谷旅游开发中，两河口有望恢复商业和公共功能，并成为当地人和游客的聚集地。该展览充分体现了典型土家干栏木构的多变性、灵活性和适应性。通过实体建筑体系的改造，古街在不同功能之间切换，公共空间从日常零售到当地集市、节庆庆典等发生拓扑变形。

The exhibition demonstrates the renovation of Lianghekou Village, which is part of a general masterplan for the entire Tujia valley, along with the ancient Salt Road. In the upcoming tourism development of the valley, Lianghekou Village is expected to resume its commercial and public functions and to be revitalized as a gathering place for locals and tourists. The project makes full use of the variability, flexibility and adaptability of the typical of Tujia Ganlan wooden structure. Through the transformation of the physical construction system, the ancient street is switched among different programs with topological deformations of public space from daily retailing to local markets and festival celebrations.

东亚木构文化圈中的干栏木构建筑遗产代表
Representative of Ganlan wooden architectural heritage in the east Asian wooden structure cultural circle

上：韩国河回村屏山书院对晚楼
Up: Duiwan pavilion in Heho Village in South Korea
图片来源：张彤摄
中：中国贵州西江千户苗寨
Middel: Qianhu Miao Village in Xijiang, Guizhou, China
图片来源：https://www.istockphoto.com
下：日本京都桂离宫
Below: Katsura Imperial Villa in Kyoto, Japan
图片来源：《石元泰博写真集 · 桂離宮 · 空間と形》

5.2 展览内容
Exhibition Content

● 干栏
Ganlan

干栏，是一种与亚洲照叶林带温湿气候、稻作经济及席居生活相适配的建造文化系统，广泛分布于日本列岛、朝鲜半岛、中国南方各省、东南亚中南半岛及其他诸国。干栏建筑表现为主要由木构件构成的建造体系和空间形态，其开放交互的围护界面适应湿热的气候，其架离地表的高床火铺适应崎岖的地形。干栏木构的构件加工和建造工法体现出高度发展的体系性和装配化，使其得以超越一时一地的需求和条件，显现出可变性、灵活性和适应性。

Ganlan (Pile Timber Construction) is a building-culture system that adapts to the warm and humid climate in the Asian leafy forest belt, the economy of rice cultivation, and the lifestyle of seating dwelling, extensively distributing in the Japanese archipelago, the Korean Peninsula, various provinces in southern China, Southeast Asia Indochinese Peninsula and other countries. The construction system and spatial form of Ganlan architetures are mainly composed of wooden components. Its open and interactive building envelope is adapted to the humid tropical climate, and its elevated floor is adapted to rugged mountainous terrain. The highly developed modular system and assembly construction of Ganlan exceeds the given conditions of a specific program, site, and time and demonstrates its variability, flexibility and adaptivity.

● 川盐济楚 · 古盐道

Chuanyan Ji Chu · the ancient Salt Road

“川盐济楚”是晚清时期太平天国军在兴起以后，因淮盐不能运到湖北而采用的政策。川盐所“济”的是宜昌府、施南府、鹤峰厅。“川盐济楚”是中国近代史上的重大历史事件。

The policy of "*Chuanyan Ji Chu*" was adopted during the late Qing Dynasty after the Taiping Heavenly Kingdom's military was on the rise, as Huai salt could not be transported to Hubei. The Sichuan salt "assists" Yichang Prefecture, Shinan Prefecture, and Hefeng Hall. "*Chuanyan Ji Chu*" is a significant historical event in modern Chinese history.

四川盐井
Sichuan salt wells

“川盐济楚”：在鄂西地区的古盐道路线
"*Chuanyan Ji Chu*": the ancient Salt Road route in the western Hubei region

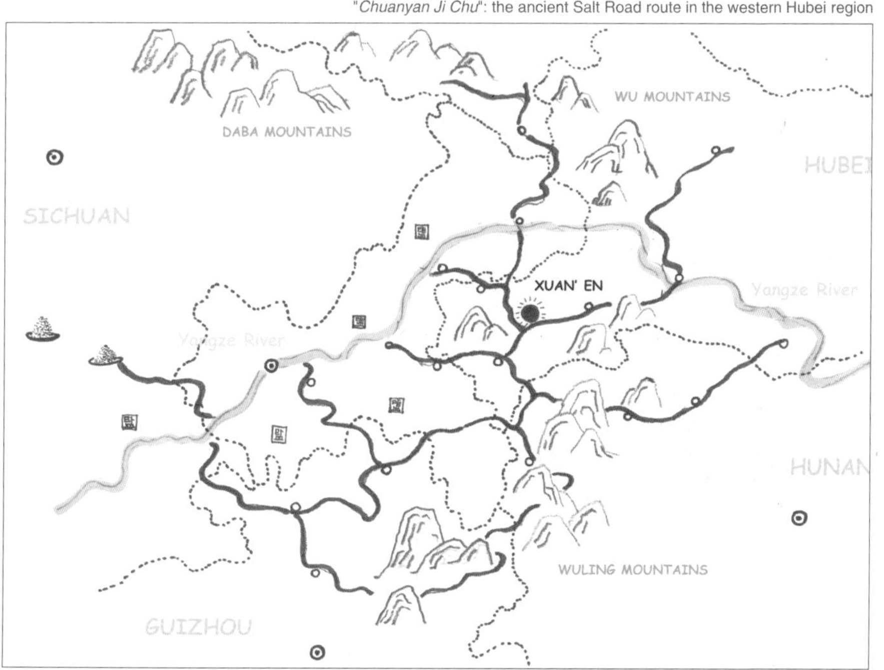

两河口老街
Lianghekou old street

两河口老街区位
Location of the Lianghekou old street

图片来源：Zhang Tong, Xu Han, Wang Chuan. Self-adaptability and topological deformation of Ganlan architectural heritage: conservation and regeneration of Lianghekou Tujia village in Western Hubei, China[J]. Frontiers of Architectural Research, 2022, 11 (5) : 865-876.

● 两河口老街
Lianghekou old street

湖北省恩施土家族苗族自治州宣恩县沙道沟镇位于酉水河上游，历史上是“川盐济楚”古盐道的重要枢纽。位于沙道沟镇的彭家寨和两河口老街，是中国中部武陵山区保存最为完好的传统聚落遗存，也是亚洲干栏木构建筑遗产的典型样本。

Located in the upper reaches of Youshui River, Shadaogou Town, Xuan'en County, Enshi Tujia and Miao Autonomous Prefecture, Hubei Province, historically served as an important hub of the ancient Salt Road of "*Chuanyan Ji Chu*". Pengjia Village and Lianghekou old street in Shadaogou Town are the best-preserved relics of traditional settlements in the Wuling Mountains in central China, and they are also typical samples of Asian Ganlan wooden architectural heritage.

两河口老街三维扫描测绘

3D-scanning measurement survey of the Lianghekou old street

图片来源：Zhang Tong, Xu Han, Wang Chuan. Self-adaptability and topological deformation of Ganlan architectural heritage: conservation and regeneration of Lianghekou Tujia village in Western Hubei, China[J]. Frontiers of Architectural Research, 2022, 11 (5) : 865-876.

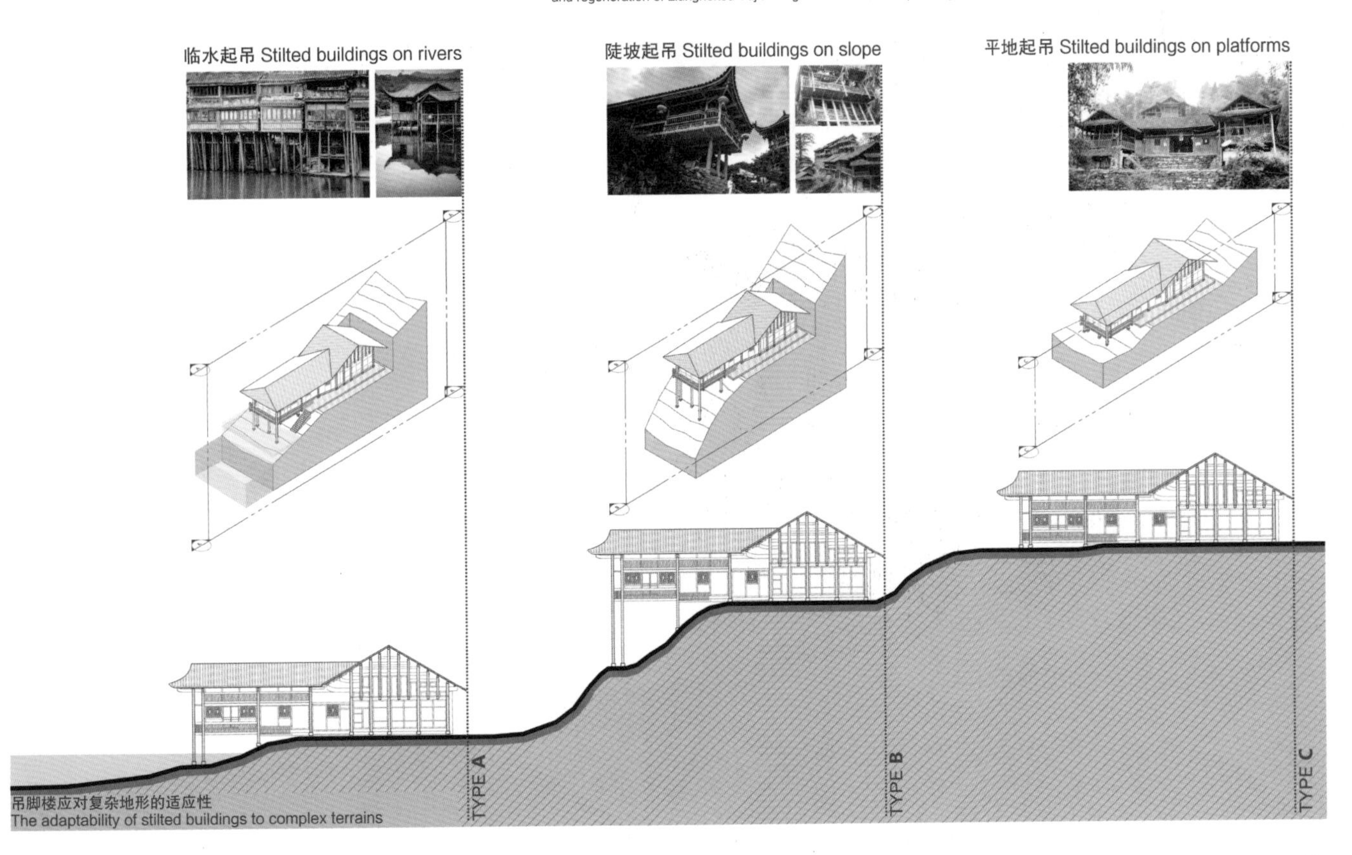

吊脚楼应对复杂地形的适应性

The adaptability of stilted buildings to complex terrains

两河口河谷聚落模型
Model of the Lianghekou river valley settlement

土家族传统聚落中村落环境和吊脚楼模型
Models of village environment and stilted buildings in Tujia traditional settlement

土家族传统吊脚楼模型 1
Models of Tujia traditional stilted buildings 1

土家族传统吊脚楼模型 2
Models of Tujia traditional stilted buildings 2

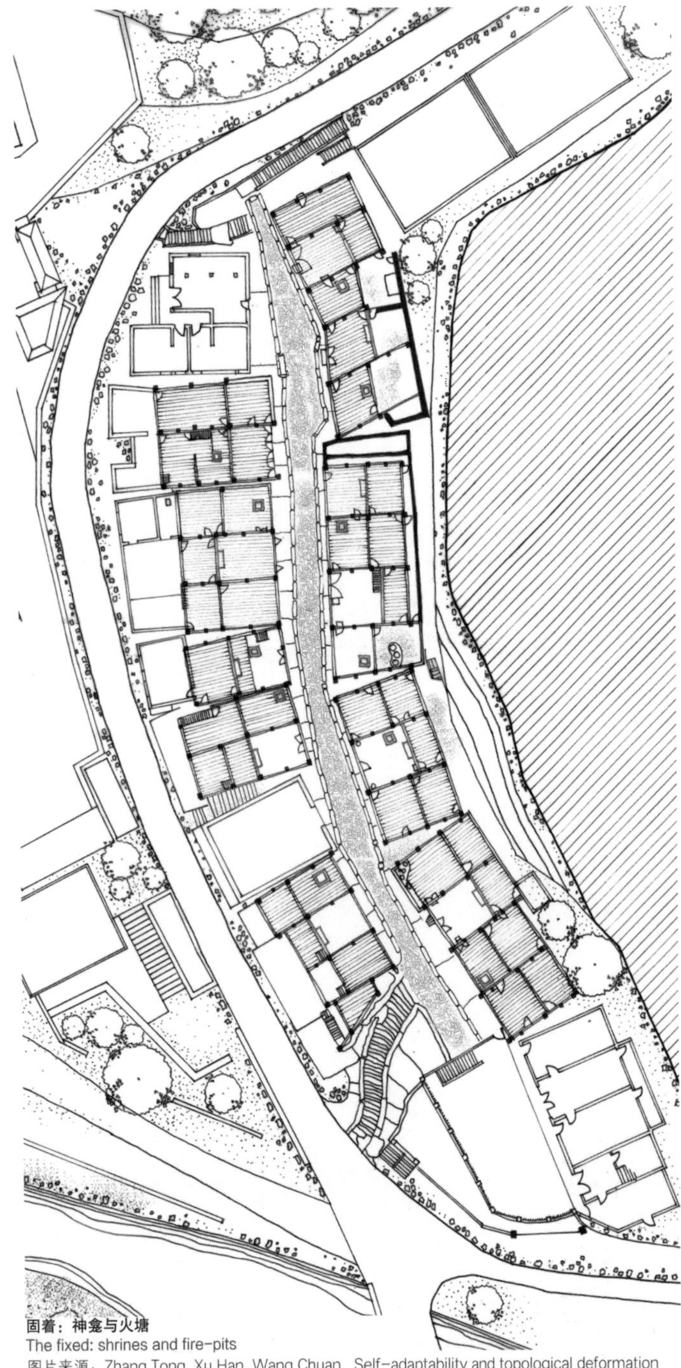

固着：神龛与火塘

The fixed: shrines and fire-pits

图片来源：Zhang Tong, Xu Han, Wang Chuan. Self-adaptability and topological deformation of Ganlan architectural heritage: conservation and regeneration of Lianghekou Tujia village in Western Hubei, China[J]. Frontiers of Architectural Research, 2022, 11 (5) : 865-876.

● 固着与可变的建造体系

Fixed and variable archi-tectonic systems

两河口老街是“川盐济楚”古盐道网络中的重要节点，其百米长的连续穿斗木构排架，构成了典型的线性干栏聚落。根据生活形态和建造方式，老街的建造体系可分为固着和可变两个子系统。固着系统，包括10座神龛和10个火塘以及部分砖石构筑；所有的木质围护结构都可归入可变系统，必要时木构排架和木楼板也可参与到可变系统中。这种高度可变和灵活的建造体系，使村寨在近200年的历史中保有居住和经济的活力，并处于稳定连续的自我更新过程中。最近20年，因人口向城市迁徙，老街逐渐废置凋敝。

土家族火塘

Tujia fire-pit

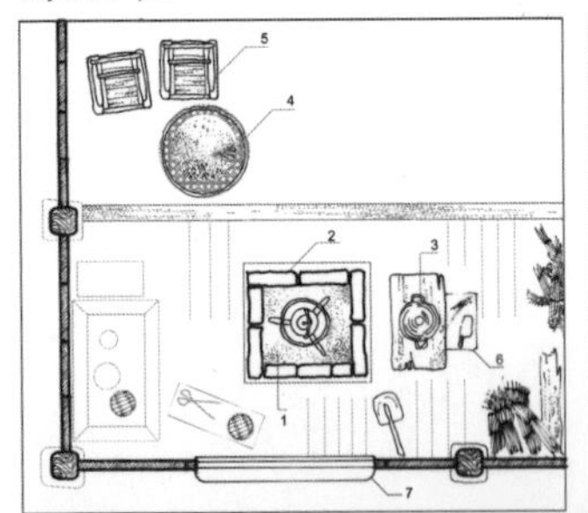

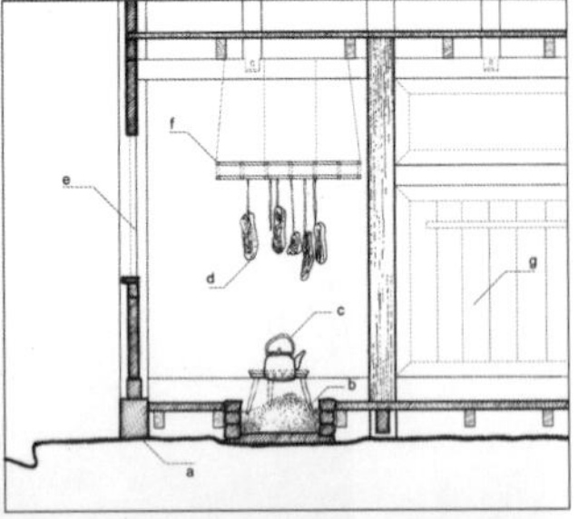

● 固着的建筑要素
The fixed architectrual elements

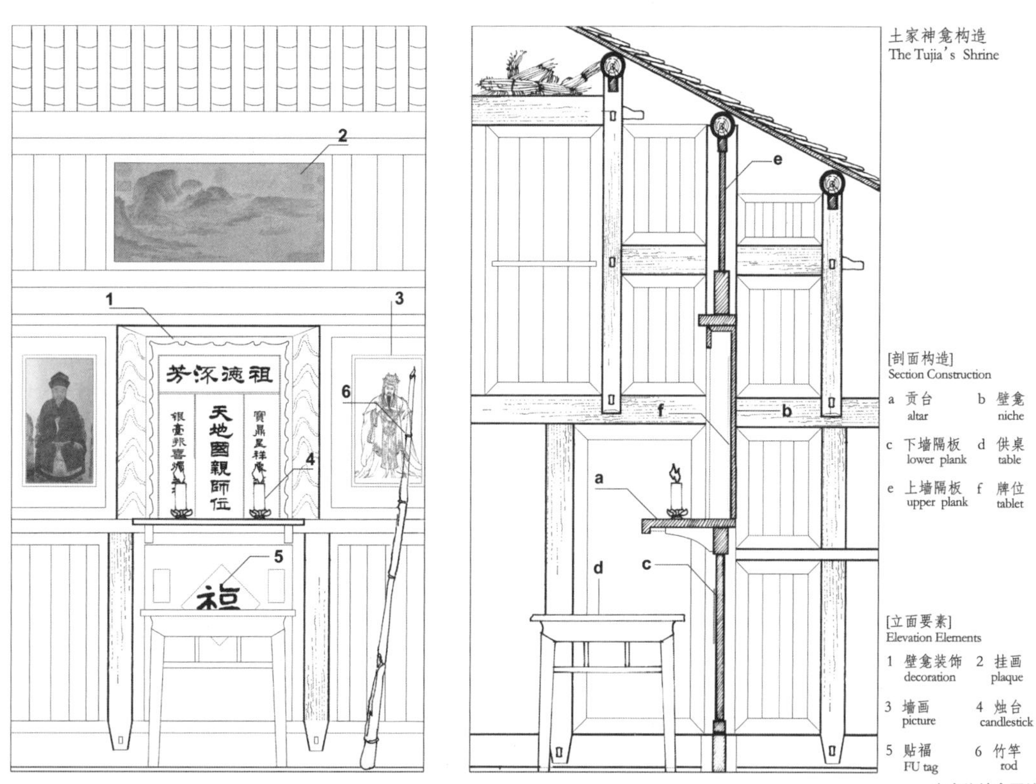

土家族神龛图绘
Drawing of Tujia shrine

图片来源：Zhang Tong, Xu Han, Wang Chuan. Self-adaptability and topological deformation of Ganlan architectural heritage: conservation and regeneration of Lianghekou Tujia village in Western Hubei, China[J]. Frontiers of Architectural Research, 2022, 11 (5) : 865-876.

土家族神龛
Tujia Shrines

● 可变的建筑要素
The variable architectrual elements

土家族干栏木构体系
Tujia Ganlan wooden structure system
图片来源：Zhang Tong, Xu Han, Wang Chuan. Self-adaptability and topological deformation of Ganlan architectural heritage: conservation and regeneration of Lianghekou Tujia village in Western Hubei, China[J]. Frontiers of Architectural Research, 2022, 11 (5) : 865-876.

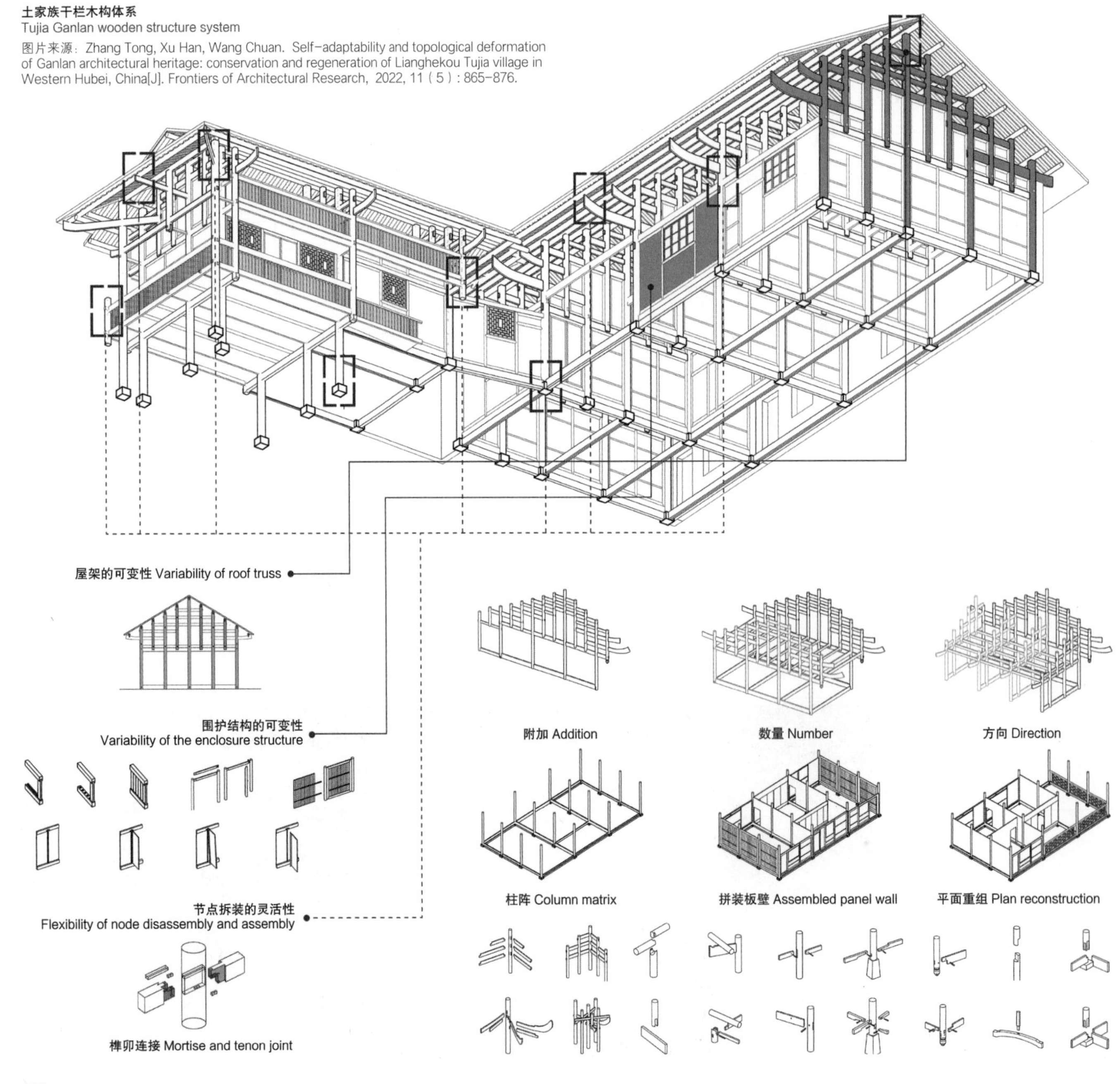

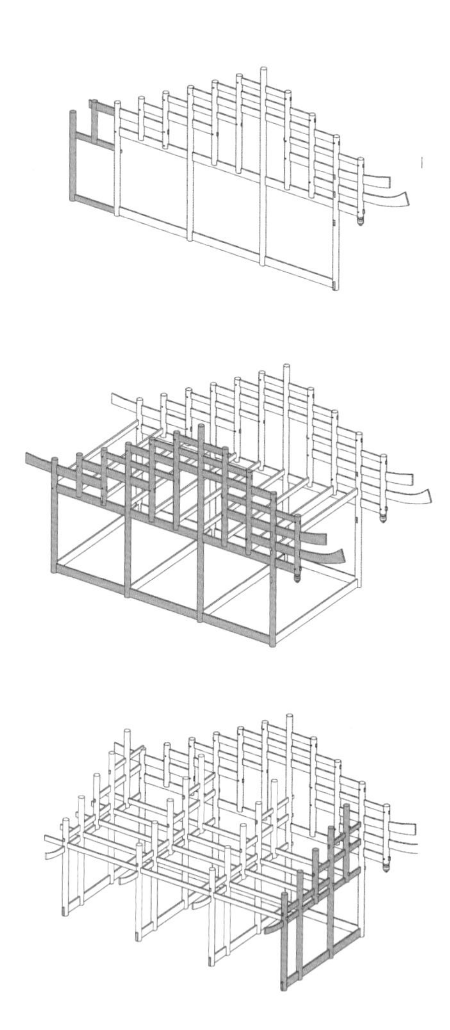

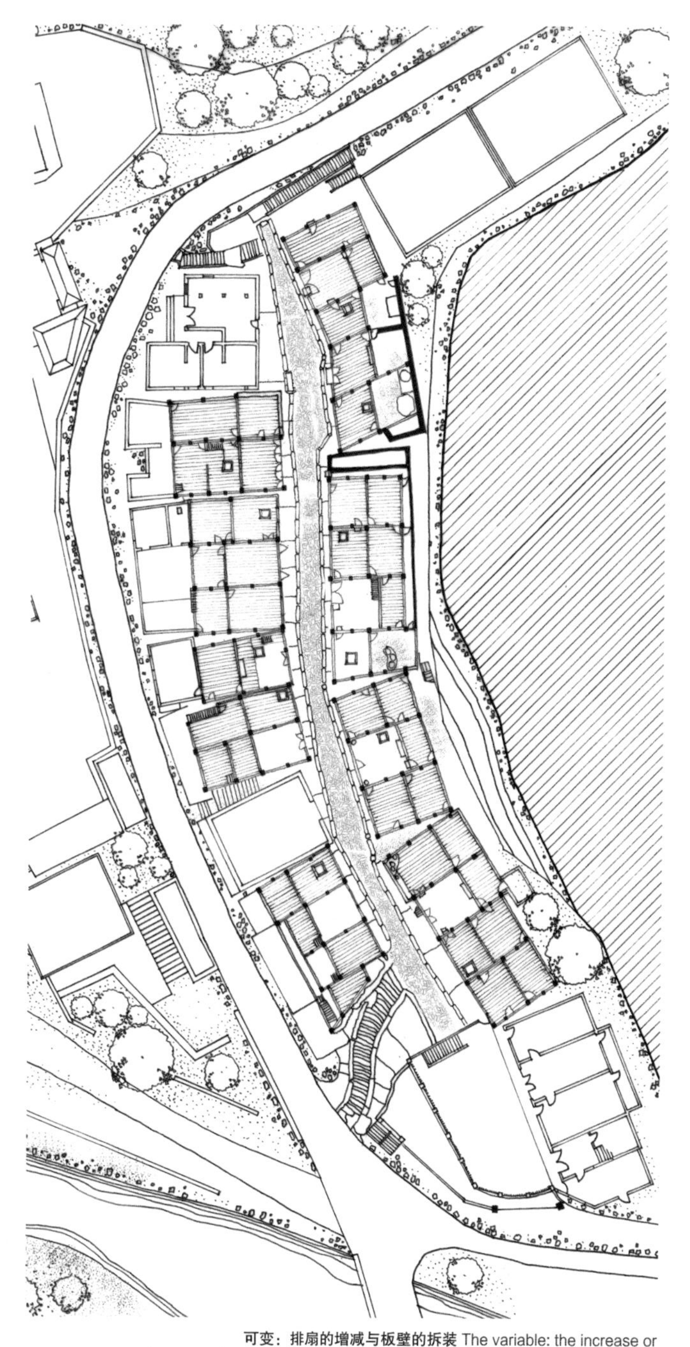

The Lianghekou old street is an important node in the ancient Salt Road network of "*Chuanyan Ji Chu*", with a hundred meter long continuous *Chuandou* wooden structure frame, forming a typical linear Ganlan settlement. According to the living style and construction method, the tectonic system of the old street can be classified into a fixed subsystem and a variable subsystem. The fixed subsystem includes ten shrines, ten fire-pits and parts of masonry; all the wooden enclosures can be regarded as a variable subsystem. Furthermore, structural timber frames and timber floors can also become a part of the variable subsystem when necessary. This highly variable and flexible construction system has enabled the village to maintain residential and economic vitality for nearly 200 years of history, and to be in a stable and continuous process of self renewal. In the last 20 years, fue to the migration of people to cities, the old street has gradually been abandoned and delayed.

可变：排扇的增减与板壁的拆装 The variable: the increase or decrease of exhaust fans and the disassembly and assembly of panel walls

图片来源：Zhang Tong, Xu Han, Wang Chuan. Self-adaptability and topological deformation of Ganlan architectural heritage: conservation and regeneration of Lianghekou Tujia village in Western Hubei, China[J]. Frontiers of Architectural Research, 2022, 11（5）: 865-876.

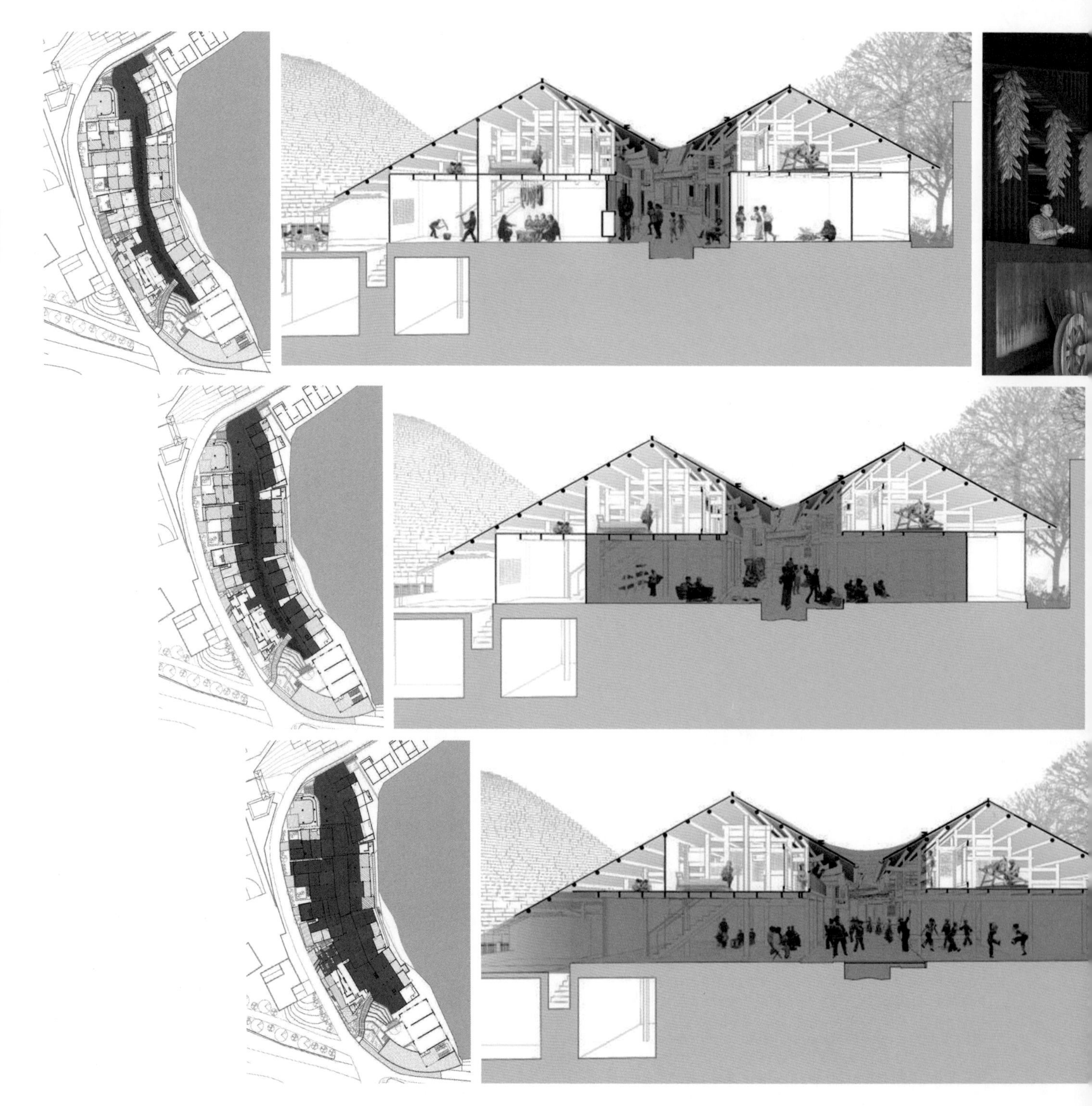

● 公共空间拓扑

Topology of public space

在盐道山谷作为土家泛博物馆实施旅游开发的背景下，两河口老街将有机会恢复其盐道贸易时期的商业服务功能，吸引村民返乡创业，并有望重新成为当地人和游客的相聚共处之所。项目设计充分利用干栏木构的可变性、灵活性和适应性，通过对老街可变要素的装、拆、移、变，实现老街公共空间在日常零售、市集赶场和节日庆典之间的切换，以同一建造体系实现多种公共空间的拓扑转化。

With the upcoming tourism development of the whole valley as a Tujia Pan-museum, Lianghekou Village will have the opportunity to resume its commercial function during the period of the Salt Road trading, attract villagers to return to their hometown and start businesses and is expected to become a gathering place for locals and tourists. The design of this project makes full use of the flexibility of Ganlan wooden structure. Through installation, demolition and transformation, the functional space can be switched among daily retailing, regular local market and festival celebration venue. With the same building system, it can achieve the topological deformation of multiple types of public space.

上：日常的老街 The ordinary old street
中：赶场的集市 The bustling market
下：庆典的摆手堂 The waving hall of the festival

● 点状激活与活态更新

选点1·大凉亭

拆除老街北端的砖混建筑，在这个地形显著变化的村口搭建一个“大凉亭”。“大凉亭”传承土家干栏木构的传统建造技艺，在村口为乡民、游客和邻近小学的孩子提供一个共处共享的公共空间。

Point activation and active renewal

Selected building 1 – Grand Liangting

At the northern entrance of the village, where the terrain significantly changes, the concrete building is planned to be demolished and replaced by a "Grand Liangting" (Chinese Loggia). Inheriting the traditional technics of Tujia Ganlan wooden structure, the Grand Liangting provides public space for villagers, tourists and children from the adjacent primary school to meet and share.

土家大凉亭模型 1
The model of Tujia Grand Liangting 1

土家大凉亭剖透视
Sectional perspective of Tujia Grand Liangting

土家大凉亭模型 2
The model of Tujia Grand Liangting 2

选点2 · 西兰卡普工坊

Selected building 2 · Silankarp workshop

西兰卡普是土家族代表性的织锦工艺。老街南段的砖混建筑将被拆除，与相邻的木构民居共同改造成为西兰卡普工坊。民艺工作坊兼具展示、售卖、研学、交流的多种功能，并传承这一当地最具代表性的活态文化遗产。

Silankapu is a representative weaving technique of the Tujia nationality.The masonry structure in the southern section of the old street will be demolished and transformed into the Silankapu workshop together with adjacent wooden houses. This handicraft workshop will combine multiple functions such as display, sales, research and exchange, and inherits the most representative live cultural heritage of the local area, Silankhap.

西兰卡普工坊模型外景 1
Exterior view of Silankarp workshop model 1

西兰卡普工坊模型外景 2
Exterior view of Silankarp workshop model 2

西兰卡普工坊模型剖透视
Sectional perspective of Silankarp workshop model

西兰卡普工坊模型内景
Interior of Silankarp workshop model

两河口老街再生活化方案效果图 1

两河口老街曾经是盐码头，承载着几代人的生活，但如今却只留下了封闭孤立的内巷，还有孤独的留守儿童和老人。所以我们的设计初衷是为他们创造一套改善生活的公共空间系统。在老街旧址中，虽然传统的土家族民居结构延续至今，但杂乱无章的现代建筑已经严重影响了老街居民的生活质量和街巷风貌。我们设想了两种策略来解决问题。

两河口老街再生活化方案效果图 2
Rendering of the regeneration and activation plan for Lianghekou old street 2

Lianghekou old street used to be a salt port that provided the livelihood for generations. Now, it has declined to an isolated inner lane, with only left-over children and elderly living in it. Therefore, our design aims to provide them with a public space system to improve their living standard. In the remaining site, although the traditional Tujia residential structure is preserved, chaotic modern buildings have severely affected the residents' quality of life and the ancient street's style. Therefore, we propose the following two strategies to solve the problem.

两河口老街再生活化方案总平面
Master plan of the regeneration and activation plan for Lianghekou old street

两河口老街再生活化方案效果图 3
Rendering of the regeneration and activation plan for Lianghekou old street 3

1. 对其三个空间节点进行“针灸”式的更新改造：
  （1）拆除北侧入口突兀的现代建筑，以新竹材料取代传统的竹结构；
  （2）对南侧入口现有老建筑进行保护修复，恢复土家族的建筑风貌；
  （3）让纯真的幼儿园“消隐”在山林中。

2. 将带状交通系统引入老街，且灵活适应地形条件，让当地老人找寻自己的空间。传统与现代的对话有助于启发当地居民的日常生活，让老街能再现往日的活力。

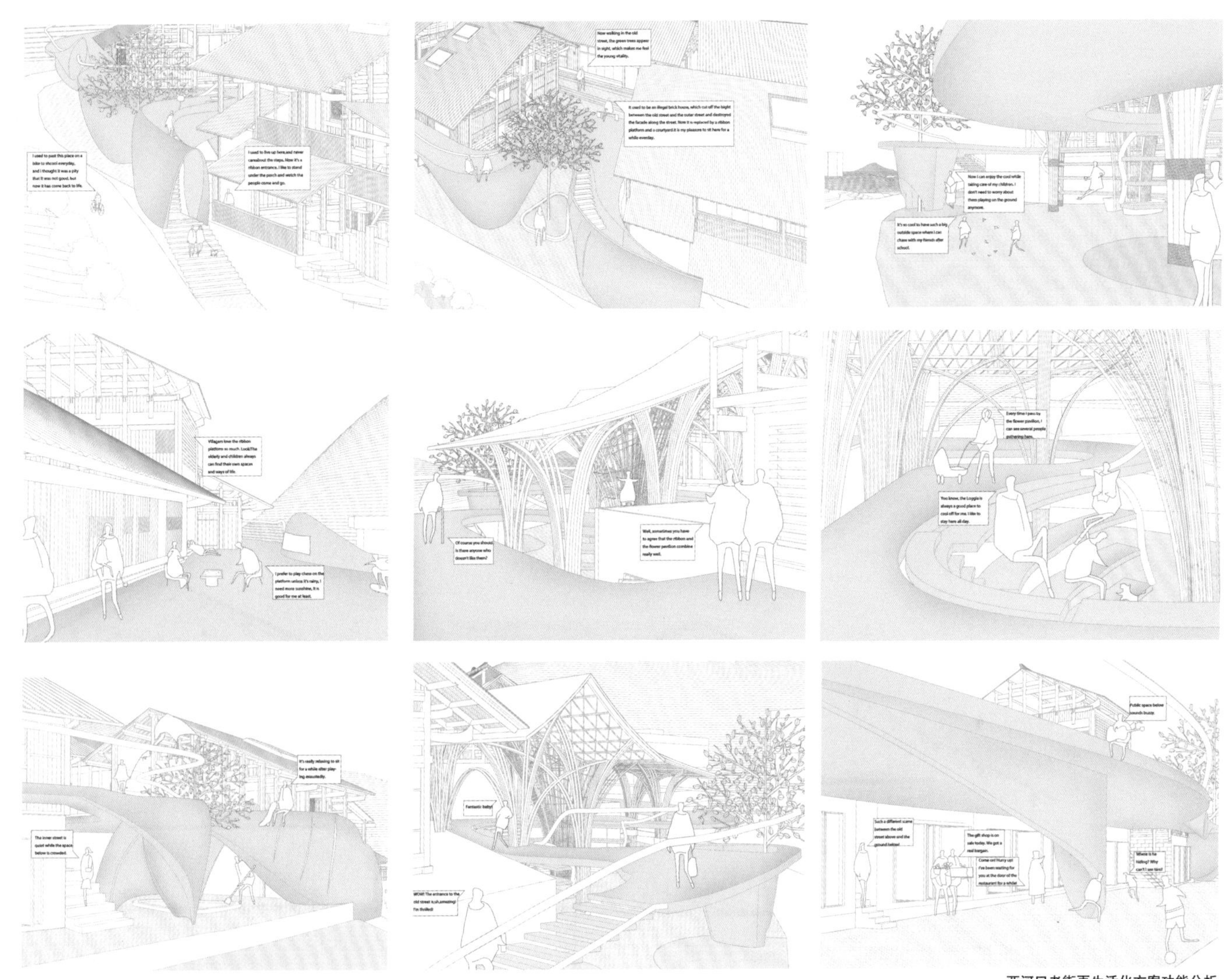

两河口老街再生活化方案功能分析

Functional analysis of the regeneration and activation plan for Lianghekou old street

1. Perform "acupuncture and moxibustion" type renovation and transformation of its three spatial nodes:
   (1) Demolish the disharmonic modern building at the north entrance, and replace the traditional bamboo structure with new bamboo materials;
   (2) Restore and protect the existing old architecture at the south entrance to revive the Tujia architectural style;
   (3) "Hide" modern kindergartens in the mountain scene.

2. Introduce a belt-shaped transportation system into the ancient street and flexibly adapt it to the terrain conditions to preserve space for the local elderly. Activate the daily life of local residents to support the revival of the ancient street through dialogue between tradition and modernity.

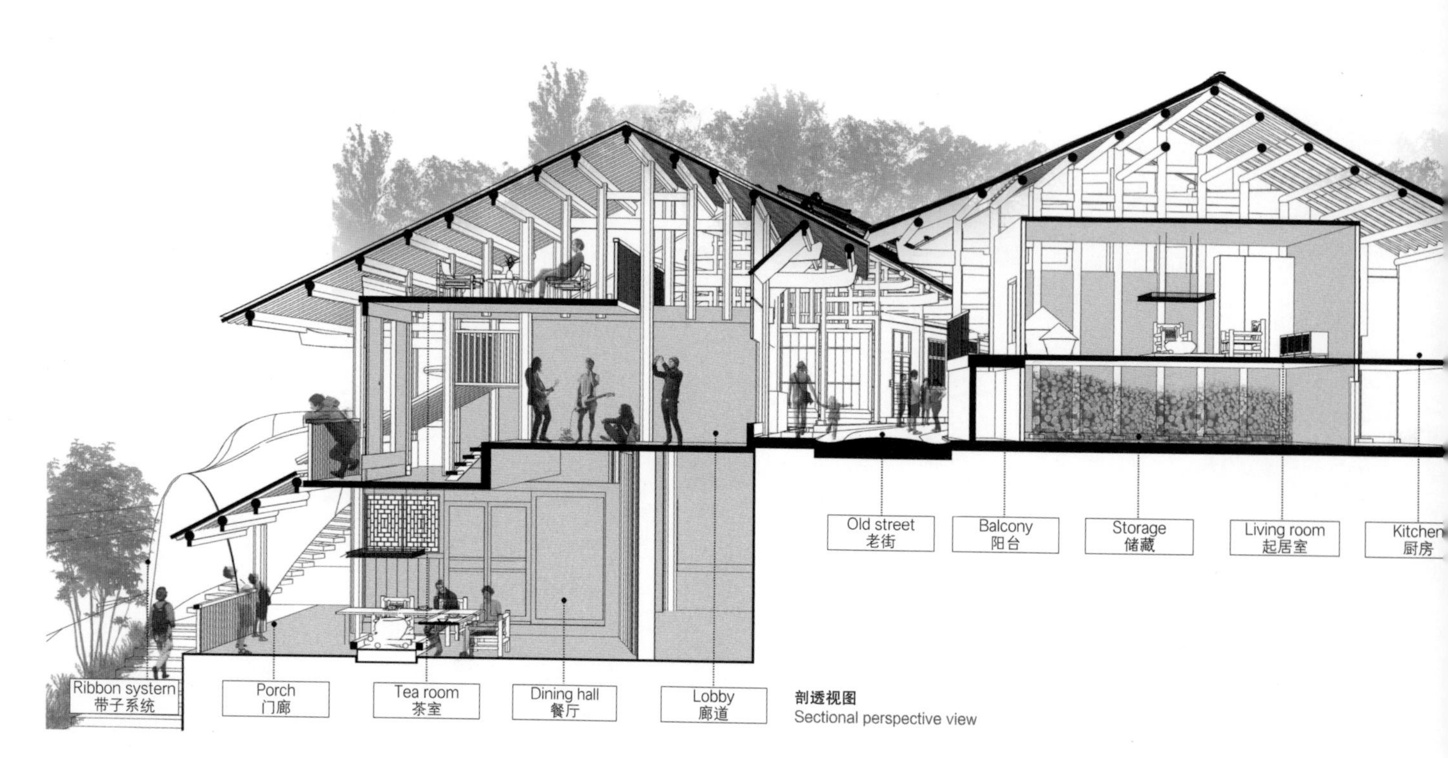

剖透视图
Sectional perspective view

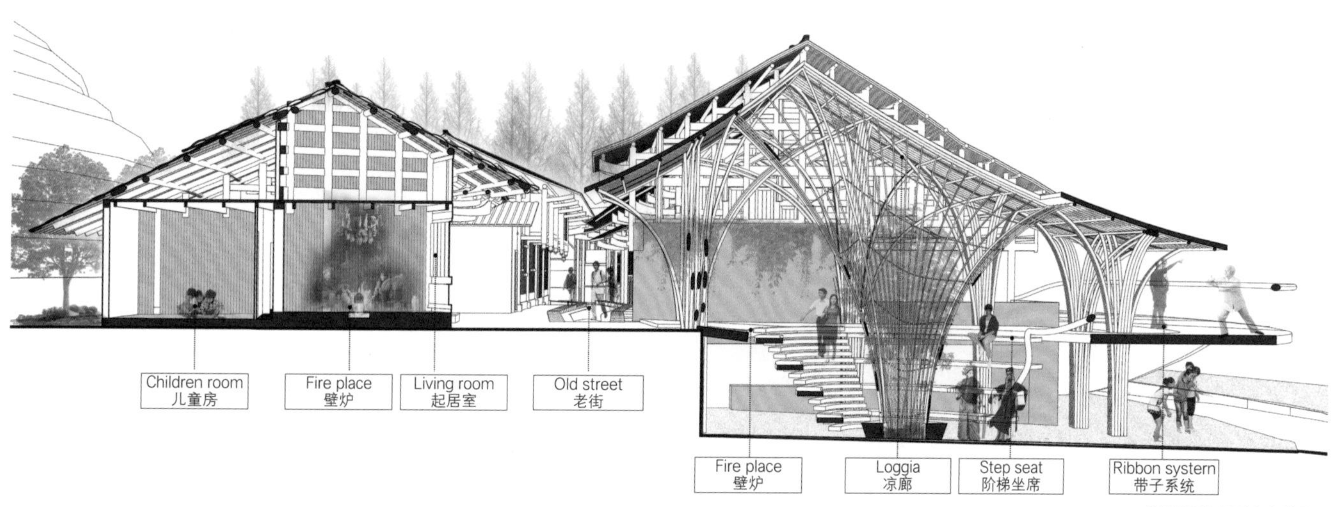

威尼斯双年展重庆大学展品
Exhibit of La Biennale in Venice by CQU

门廊空间
Porch space

两河口村的振兴和再生向我们传递了一个重要信息，即传统住房的潜在价值对我们当代住房发展非常重要。它们包含了许多潜在的元素，并具有超越现代生活的优越特性。一系列的设计方案旨在恢复那些被现代人遗忘在传统住宅中的价值和意义。两河口村的活态更新向大家展示了一种我们的城市不再能够为我们提供的共享生活的可能性。

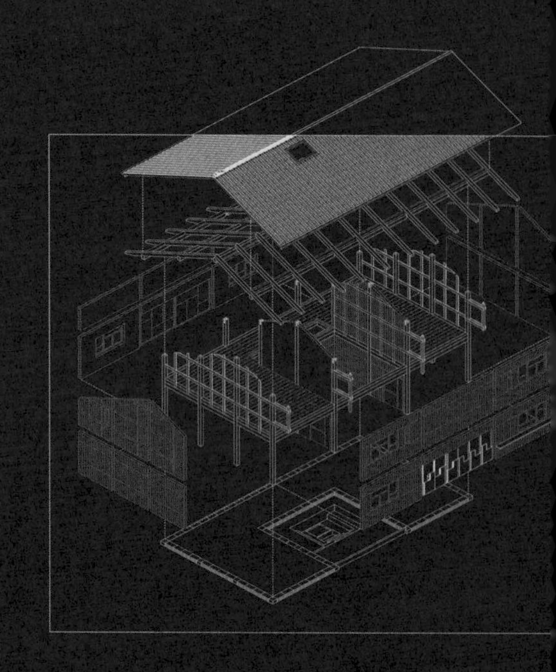

The revitalization and regeneration of Lianghekou Village has conveyed an important message to us, which is that the potential value of traditional housing is very important for our contemporary housing development. They contain many potential elements and possess superior characteristics beyond modern living. A series of design proposals are aimed at restoring the value and significance of those parts that have been forgotten by modern people in traditional housing. The active renewal of Lianghekou Village has shown everyone a possibility of shared living that our city is no longer able to provide for us.

RE-ENVISIONING LIANGHEKOU DEALING WITH TRADITION
重新展望两河口，直面传统

宣恩壹号粮仓外景 External view of Xuan'en No.1 granary
图片来源：陈颢摄

● 宣恩壹号粮仓
Xuan´en No.1 Granary

宣恩壹号粮仓建于1951年，保存完好。建筑共两层，近1 630平方米。设计借鉴苏式仓，并对传统粮仓加以改进，通风、防潮效果更好。主体为木结构，吸取当地土家族传统民居穿斗木构之精髓，以仓廒本身木构为辅助支撑，并整体抬升架空，结合西式屋架与气楼式天窗，形成颇具特色的“轿顶仓”。外围护结构为厚重的石构墙体，窗洞外设可调节遮阳板。12廒木构仓体布置于仓房内部，呈中轴对称的“品”字形分布，每组4廒。另有楼梯若干。中厅高而窄，光由侧高窗射入，塑造出类似教堂空间氛围的神圣感，具有改造为博物馆建筑的空间潜力。

The Xuan'en No.1 granary was built in 1951 and is well-preserved. It has two floors, about 1,630 square meters. The design was inspired by Soviet-style granaries, and improved the traditional granaries, resulting in better natural ventilation and moisture-proofing. The main body is a wooden structure, which absorbs the essence of *Chuandou* wooden structure of local Tujia traditional residential buildings, and uses the wooden structure of the granaries itself as auxiliary support and lifts from the ground, combining Western-style roof trusses and skylights, forms a distinctive roof like "sedan chair". The exterior envelope is a thick stone wall, with an adjustable sunshade outside the windows. There are twelve wooden structure cymbals inside, distributed in a "品" shape symmetrically. Each group has four cymbals and there are several stairs. The central hall is high and narrow, and the light descends from the high windows, creating a sacred sense similar to the sacredness of the church's spatial atmosphere, hence it has the spatial potential to transform into a museum.

右：仓廒
Right: Granary
图片来源：陈颢摄
左上：木构与空间
Top left: Wooden structure & space
左中：村民访谈
Middle left: Interview the local villagers
左下：现场调研小组
Below left: Site investigation and research group

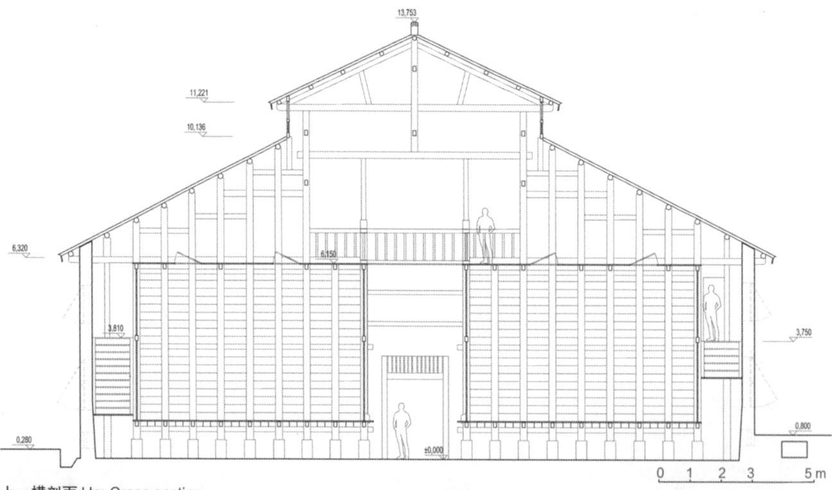

上：横剖面 Up: Cross section
下：桁架内景 Below: Interiors with the truss
图片来源：陈颢摄

原生态的通风设备
Original ecological ventilation equipment

承载上部仓廒的柱础和柱子
Foundation and the pillar of overhead lifted Granaries
图片来源：陈颢摄

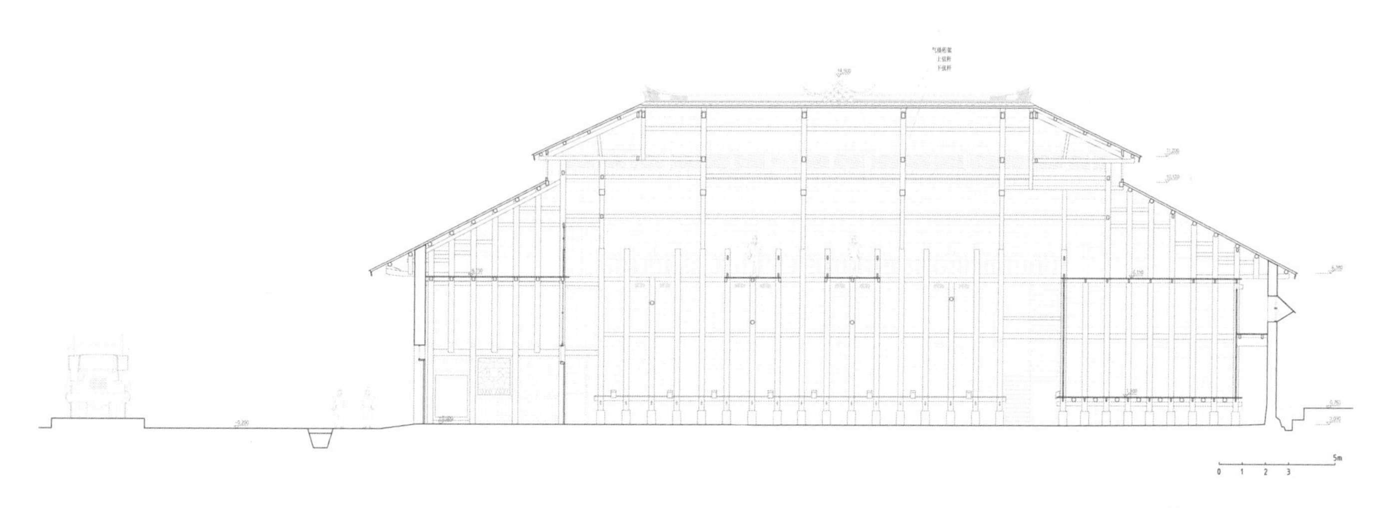

纵剖面 Longitudinal section

● 土家族风雨桥
Tujia corridor timber bridge

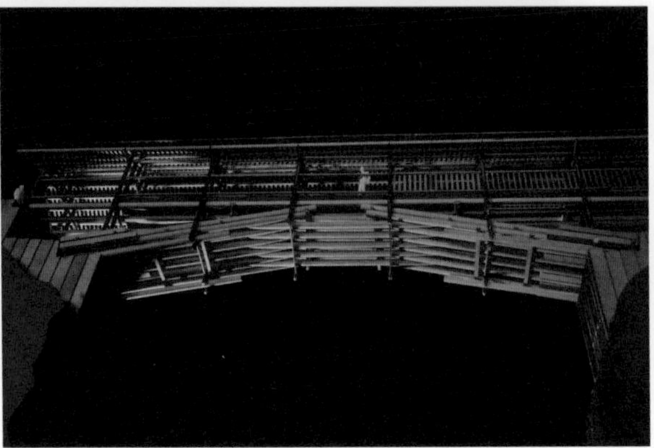

土家族风雨桥
Tujia corridor timber bridge

5.3 展览思考
Exhibition Reflection

● 探索 | 从两河口中学习

威尼斯建筑大学及三所中国大学（东南大学、华中科技大学、重庆大学）的师生以两河口古村为案例展开联合教学。通过研讨会和调研，师生们了解了两河口型形态结构，以及它与环境和整个山谷的传统关系，有些可能延续至今，而另一些直接关系在过去几十年中或已中断。两河口的现状表明，其竞争力和吸引力正在逐渐下降。本地人口减少及迁出、地理意义和社会意义上的边缘化也凸显了同样的情况。与之形成鲜明对比的是，中国中央和地方政府在将土家古村视为国家级文化遗产的"第一"考量下，采取了复兴改造的方案、政策和行动，并为原本几乎无法抵达的地区建造了高速公路等基础设施。

核心问题是，如何在新基础设施实现的必要性与特定景观保护、场所保护、传统建筑紧急修复的必要性之间形成正向的平衡，以及创造能够振兴区域、吸引人们重新聚居的规模经济。由此，两河口变成了一个项目实验室，可在此探索或许能在其他条件类似的村庄（根据当地实际情况适当调整后）效仿的策略。

小型村庄的生态可持续系统

可持续的再生或更新策略离不开对整个土家族山谷系统的全面研究，也离不开对小型传统村庄复杂系统的把控。两河口原本是商贸集镇，直线距离有限（仅不到 100 米）。将其作为第一个可通过现有基础设施抵达的村庄，作为整个山谷的"新门户"，作为首先被调研的遗址，是跟踪了解本地具体特点的一个途径。作为紧急保护建筑和景观项目的切入点，两河口极具案例研究价值。

两河口有传统建筑、传统建筑技术以及反复出现的形态元素。但是，原本用于聚居生活的空间已大部分被改变或缩小，新的加建和改建也改变了吊脚楼原有的特征。建筑与区域丘陵景观、土壤的农业用途、河流的存在以及盐路的存在的直接关系（实际上是中断的），仍然是可识别的连接元素。

Exploring | Learning from Lianghekou

The traditional village of Lianghekou has been used as a study case of the joint on-site workshop, attended by professors and students of IUAV and of three Chinese universities, i.e. SEU, HUST, CQU. The seminar and the surveys made it possible to understand the Lianghekou type-morphological structure and the traditional relationships (possibly, still existing) with its surroundings and with the entire valley. Some of those direct relationships have been interrupted through the last decades. The actual conditions show a progressive lack of competitiveness and attractiveness of the places. These same conditions are underlined by the progressive depopulation and abandonment and by physical and social marginalization phenomena. In contrast to these dynamics, the Chinese center government and local authorities have undertaken alternative regeneration and renovation programs, policies and actions, supported by the "first" consideration of Tujia villages as a part of the national heritage, as well as, have started the construction of new infrastructures (i.e., a new highway) to serve almost not accessible lands portions.

Therefore, the central question is how to create a positive metastasis between the necessities of new infrastructure realizations and those of specific landscape protections, of places conservations, of traditional architectural emergencies restorations; and how to create scalar economies which allow to revitalize, to rehabilitate places. This is how Lianghekou becomes a laboratory, a place to experiment possibly replicable project strategies (with site-specific appropriate adaptations) in other villages with similar conditions.

穿越土家族山谷、连接传统村庄的古盐道（应结合土地基础设施和河道，重新考虑盐道的名称）在几个世纪以来一直是最重要的基础设施。要将现代城市的条件带给两河口，意味着需要赋予这一传统基础设施新的价值，为创造土家族村庄，尤其是位于同一个山谷的村庄之间新的联系建立新情景，还意味着需要重新激活盐道曾经建立并支持的复杂关系系统。纵观整个小村庄体系，我们了解并注意到小村庄在位置、规模、内部特征、与周边地区关系等方面的具体作用，明确具体的生活方式和使用场所，并从这些场所入手，寻找可持续的发展战略。

与传统的关系

两河口所有历史建筑都受到了传统吊脚楼的影响。一定程度上，最早的吊脚楼类型受到了在独特空间内修复住宅和商业空间的必要性的影响。这种独特的特征，即生活和商业空间混合在一起的情况，在与其他村庄建筑相比时尤其明显：在其他村庄的建筑中，居住的部分更为发达，且仅从建筑类型、内部空间布局和是否面对盐道等方面便可轻松辨认。

两河口允许对现有建筑进行保护和创新进行可能性探索，这些建筑或多或少受到了过去几十年来发生的深刻变化的影响。两河口有完全使用混凝土建造的建筑，有保存完好但急需维护和修复的传统建筑，还有传统结构经过严重改造、带有劣质加建部分的建筑。这改变了内部和外部的关系、建筑物与地面接触的方式、内部和外部空间的顺序，对生活质量至关重要的过渡空间也被转变成了外部空间。以上所有条件，都为自由设计练习创造了机会。

（作者：朱塞佩·卡尔达罗拉）

An eco-sustainable system of small villages

Sustainable redevelopment or renewal strategies cannot be separated from a comprehensive approach to the entire system of the Tujia valley and dealing with the complex system of its small traditional villages. With its original commercial character and vocation and with its limited linear shaped configuration (less than 100 m), Lianghekou acquires case study values asitis considered the first village that can be reached through the existing infrastructures, a new "door" for the entire valley, a privileged and primarily surveyed site in which to trace and understand the specific characters of the places and a project opportunity to operate in terms of conservation of the architectural and landscape emergencies.

There are traditional architectures, with their construction techniques, and recurring type-morphological elements, but there are many facts that have principally altered (and reduced) spaces originally dedicated to collective living and there are buildings additions that have modified the original characteristics of the stilted buildings. There are direct relationships (actually, interrupted) with the regional hilly landscape, with the agricultural uses of the soils, with the presence of the river and precisely of the Salt Road which are still recognizable as connective elements.

The ancient Salt Road—more correctly, under the name of Salt Road should be reconsidered by combining the land infrastructure and the river way—crosses the Tujia valley and connects the traditional villages; it has been an infrastructure of primary importance over the centuries. Establishing new urban conditions for Lianghekou means assigning new values to this traditional infrastructure, generating a new scenario for activating new possible relationships between the villages of the Tujia nationality, especially between those located in the same valley and also means reactivating that complex system of relationships that proper the Salt Road was able to establish and support. Looking at the whole system of small villages allows us to understand and highlight specific roles in terms of location, size, internal characteristics, and relationships with the surrounding areas, it allows us to clarify specific ways of living and using places and, starting from them, to identify sustainable development strategies.

Dealing with tradition

All the historic buildings in Lianghekou are inspired by the traditional stilted buildings. The original typology is partially affected by the necessity of recovering, within a unique perimeter, residential and commercial spaces. This specific condition—a sort of hybrid character (living and commercial spaces joined together) emerges in comparison with the other village buildings, where the residential vocation was more developed, and easily readable just by looking at the building types and the internal spatial distribution or to the facades along the Salt Road.

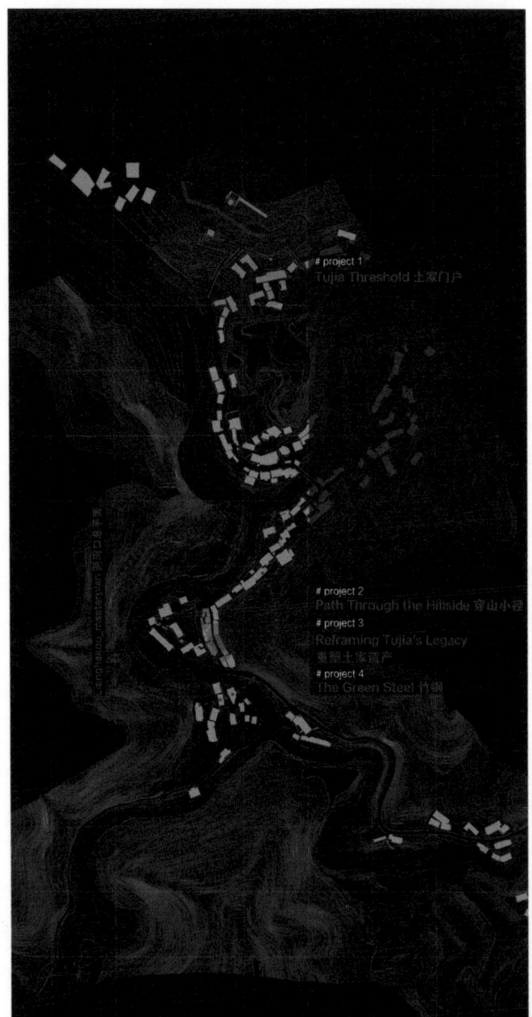

Lianghekou allows to experiment alternating possibilities of conservation/innovation of the existing buildings; these buildings have been affected by more or less profound changes through the last decades. There are buildings entirely in concrete, preserved traditional buildings but in degraded conditions with a strong necessity for maintenance and restoration, and buildings with low-quality additions where traditional structures have been significantly modified. This also alter for the internal-external relationships, the ways by which the buildings touch the ground, the sequences between internal and external spaces, and precisely colonizing those in-between spaces which were fundamental for the quality of living. All these specific conditions allow free design exercises.

(Author: Giuseppe Caldarola)

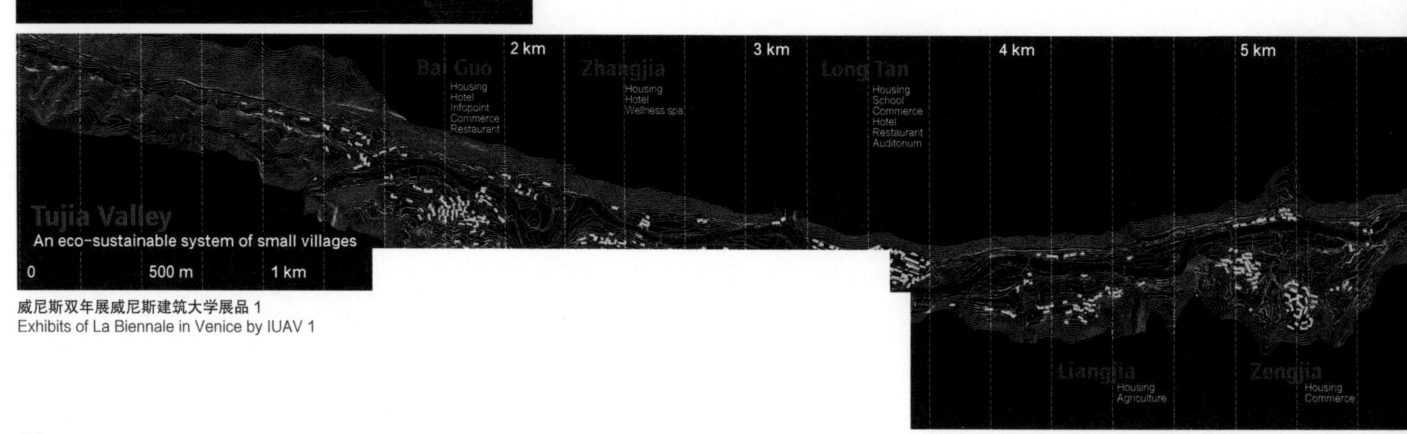

威尼斯双年展威尼斯建筑大学展品 1
Exhibits of La Biennale in Venice by IUAV 1

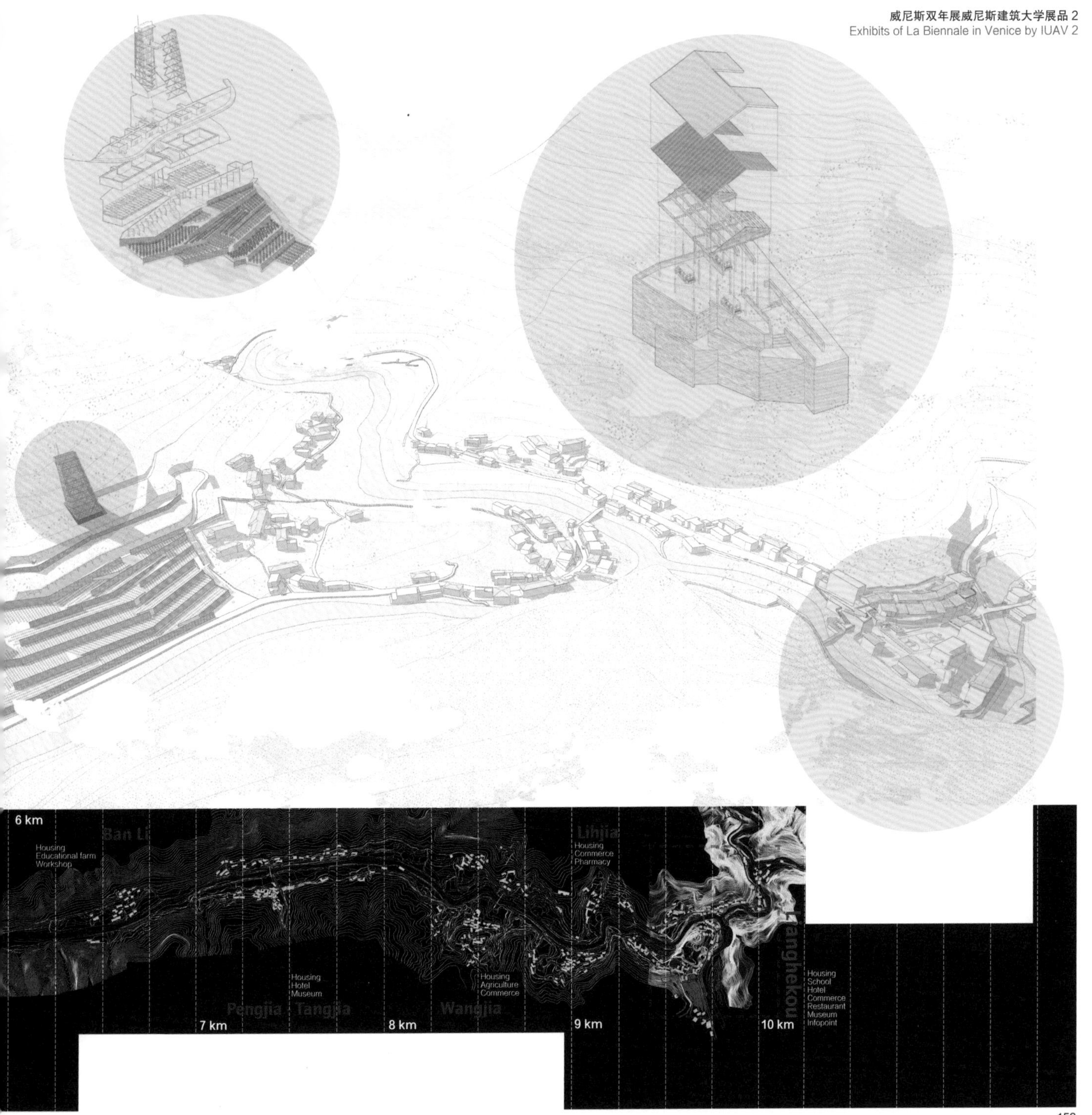
6 km
Ban Li
Housing
Educational farm
Workshop
Lihjia
Housing
Commerce
Pharmacy
Housing
Hotel
Museum
Pengjia
Tangjia
7 km
Housing
Agriculture
Commerce
Wangjia
8 km
9 km
10 km
Housing
School
Hotel
Commerce
Restaurant
Museum
Infopoint

土家吊脚楼的典型特征构造“罳檐” *Siyan,* typical characteristic and construction of Tujia stilted buildings
图片来源：任紫湫摄

● 结语：另一种传统，另一种更新

干栏，这种最早发现于中国姚江流域河姆渡遗址中的木构建造体系，以其高度发达的可变性和灵活性，广泛适应于东亚各地的经济、生活与文化。在时间进程和时代更替中，干栏木构的建成遗产不再固求由砖石建筑定义的独一性和原真性，体现出一种自适应、拓扑性的活态更新机制。以土家两河口古街为例的研究和实践，旨在求证即便当代城乡生活发生如此巨大的激变，这种别样的传统依然具备如生命更新般的自适应和可持续再生能力。

Conclusion: another tradition, another regeneration

Ganlan, the wooden construction system that was first discovered in the Hemudu site in the Yaojiang River Basin in China, is widely adapted to the economy, life and culture of East Asia with its highly developed variability and flexibility. Rather than seeking for the irreplaceability and authenticity defined by the tradition of masonry architecture, the built heritage of the Ganlan wooden structure embodies a self-adaptive and topological living mechanism of regeneration. The research and practice with the case of the Lianghekou Tujia ancient street regeneration aims to prove that even if there are such drastic changes in contemporary urban and rural life, this unique tradition still possesses the ability of self-adaptation and sustainable regeneration like natural regeneration.

策展人
Curators

张彤
Zhang Tong

恩里科·丰塔纳里
Enrico Fontanari

阿尔多·艾莫尼诺
Aldo Aymonino

参展人名单
Participating Exhibitors

东南大学
Southeast University

教师：张彤、王川、李海清

Teachers: Zhang Tong, Wang Chuan, Li Haiqing

学生：徐涵、隋明明、陈斯予、马雨萌、陈韵玄、沈洁、闫宏燕、戈世钊、续爽、王瑢、李心然、任紫湫、杨宸、王东平、朱翼、杨灵、张卓然、唐宇轩、蓝明发、郑舒欣

Students: Xu Han, Sui Mingming, Chen Siyu, Ma Yumeng, Chen Yunxuan, Shen Jie, Yan Hongyan, Ge Shizhao, Xu Shuang, Wang Rong, Li Xinran, Ren Ziqiu, Yang Chen, Wang Dongping, Zhu Yi, Yang Ling, Zhang Zhuoran, Tang Yuxuan, Lan Mingfa, Zheng Shuxin

华中科技大学
Huazhong University of Science and Technology

教师：李保峰、汤诗旷、王通

Teachers: Li Baofeng, Tang Shikuang, Wang Tong

学生：张师维、魏迪、赵蕊、王鑫琪、周卓、康雪薇、张瑞芳、何仁、焦媛媛、李杏、吕枭、杨肖、贺彬亮、王沛泽、彭欣怡、王宁、田淑颖、何仕轩、刘文玉（湖北美术学院）、王长曦（湖北美术学院）

Students: Zhang Shiwei, Wei Di, Zhao Rui, Wang Xinqi, Zhou Zhuo, Kang Xuewei, Zhang Ruifang, He Ren, Jiao Yuanyuan, Li Xing, Lü Xiao, Yang Xiao, He Binliang, Wang Peize, Peng Xinyi, Wang Ning, Tian Shuying, He Shixuan, Liu Wenyu (Hubei Institute of Fine Arts), Wang Changxi (Hubei Institute of Fine Arts)

重庆大学
Chongqing University

教师：褚冬竹、宫聪、覃琳

Teachers: Chu Dongzhu, Gong Cong, Qin Lin

学生：顾明睿、罗国力、谭泊文、李慧敏、沈丹杰、庞家丽、曾昱玮、王雨寒、郭师竹

Students: Gu Mingrui, Luo Guoli, Tan Bowen, Li Huimin, Shen Danjie, Pang Jiali, Zeng Yuwei, Wang Yuhan, Guo Shizhu

威尼斯建筑大学
Università Iuav di Venezia

教师：阿尔多·艾莫尼诺、恩里科·丰塔纳里、朱塞佩·卡尔达罗拉

Teachers: Aldo Aymonino, Enrico Fontanari, Giuseppe Caldarola

学生：乔凡娜·博尔丁、伊拉里亚·波提诺、达维德·布鲁内里、达维德·布诺、安德里亚·康西安、阿尔伯特·坎顿、埃莉萨·切诺、马克·达尔托、安娜·马尔赛拉、托马斯·奥尔托兰、贾科莫·雷托雷、伊拉里亚·洛塞伦、亚里山德罗·斯帕拉潘、阿莱格拉·泽恩

Students: Giovanna Bordin, Ilaria Bottino, Davide Bruneri, Davide Burro, Andrea Cancian, Alberto Canton, Elisa Cielo, Marco D'Altoè, Anna Marsella, Thomas Ortolan, Giacomo Rettore, Ilaria Rosolen, Alessandro Sparapan, Allegra Zen

宣恩县
Xuan'en County

邵爱华、姚元文、朱朝春、杨蕊、涂先绪、成涛、赵涛、孟友顺、杨长安、宋文、卢昌群、季建华、田忠、王坤、田星橙、邹莉华

Shao Aihua, Yao Yuanwen, Zhu Chaochun, Yang Rui, Tu Xianxu, Cheng Tao, Zhao Tao, Meng Youshun, Yang Chang'an, Song Wen, Lu Changqun, Ji Jianhua, Tian Zhong, Wang Kun, Tian Xingcheng, Zou Lihua

雅伦格文化艺术基金会
Fondazione EMGdotART

马里诺·福林、李家豪、孙鹏、米凯拉·帕科、布鲁诺·马利奥托

Marino Folin, Victor Lei, Miranda Sun, Michela Pacco, Bruno Mariotto

第六章 武陵干栏

中意国际四校联合研学营在地展

CHAPTER 6 WULING GANLAN

Sino–Italian Four Universities International Joint Studio On–site Exhibition

6.1 亚洲木构建筑的文化结晶与样本遗存展览

Splendid Archi–tectonic Legacy of Asian Wooden Architecture Exhibition

威尼斯国际建筑双年展览结束后，全部展品回到宣恩县两河口村现场，策展团队将内容进一步扩展，以“武陵干栏：亚洲木构建筑的文化结晶与样本遗存”为主题，使其成为展示亚洲干栏木构建筑遗产价值及其保护与再生策略的在地常设展览。展览开幕式于 2022 年 8 月 10 日，与“亚洲干栏木构建筑遗产的保护与再生学术研讨会”国际论坛同期举办。

武陵干栏展览现场

Site of Wuling Ganlan exhibition

图片来源：湖北省恩施土家族苗族自治州宣恩县文化旅游局提供

After the La Biennale, all exhibits were shipped back to the site of Lianghekou Village in Xuan'en County. The curation team further expanded the content, with the theme of "Wuling Ganlan: Splendid Archi-tectonic Legecy of Asian Wooden Architecture", making it a permanent local exhibition showcasing the value of Asian Ganlan wooden architectural heritage and its protection and regeneration strategies. The opening ceremony of the exhibition was held on August 10, 2022, in conjunction with the "International Symposium on Conservation and Regeneration of Asian Ganlan Wooden Architectural Heritage".

干栏木构建筑建造体系是展陈的核心内容。基于这些内容，展览提出了干栏木构建筑建造系统的自适应性特征，并在两河口村的设计方案中提出和放大。对于土家族来说，灵活地重建公共空间是一种在极端不同的生活条件下的实验设计，利用一个木制结构体系的可变性来满足土家族在两河口村的日常生活、市场和仪式模式的三种不同需求。

The archi-tectonic system of Ganlan wooden architecture is displayed as its centrepiece. Based on these contents, the exhibition reflected the self-adaptability of Ganlan wooden architecture as one of its intrinsic construction features, as put forward and amplified in a design proposal for Lianghekou Village. That design was experimental and addressed the extremely different living conditions of the Tujia people and the need to reconstruct public spaces flexibly, using the variability of tectonic wood architecture to meet the three facets of the Tujia people's needs in Lianghekou, i.e., their daily life, market, and ceremonies.

宣恩一号粮仓被选为干栏木构建筑样本结晶的优秀范例。一个精致的“西洋镜”式的粮仓剖面模型展示了建筑内部精巧的结构，以及粮仓的使用场景。这座粮仓的柱子和系杆框架采用了传统的土家族木制建筑技术，形成了一个独特的大空间来容纳 12 个储藏室。在干栏木构建筑的革新意义上，这种非典型的建筑形式提供了一个很好的例子，展示了利用传统建筑技术创造当代大空间的可能性。

The Xuan'en No.1 granary is featured as an excellent example of Ganlan wooden architectures. A delicate "Western Mirror" style granary section model showcases the intricate internal structure of the building and the usage scenarios of the granary. The pillars and tie frames of this granary adopt traditional Tujia wooden building techniques, forming a unique large space to accommodate twelve storage rooms. In terms of the innovative significance of Ganlan wooden architectures, this atypical architectural form provides a good example, demonstrating the possibility of using traditional architectural techniques to create contemporary large spaces.

从土家吊脚楼中直接回收并重新组装成特色建筑单元的旧构件，既是超大展品，也是创造展览空间的手段。展览空间氛围的基调是将展品在幽暗的环境中呈现。展厅所有背景墙都漆成黑色，大型木质展品及土家古物件器皿等均保持原态，未经翻新处理。展厅内的灯光照明系统与展览内容相适配。

Old components, which had been directly recycled from Tujia stilted buildings and reassembled into characteristic construction units, serve as oversized exhibits as well as means of creating exhibition spaces. The exhibition hall is an enclosed space, and the curator emphasised exhibiting the items in a dim atmosphere. All the background walls were painted in dark colours, and the wooden construction units and Tujia Antiquities and utensils retain their original appearance and are not renovated. The lighting system in the exhibition hall is compatible with the exhibition content.

这部分展览通过重建火塘创造了令人印象深刻的空间体验。在过去类似的展览中，火塘仅仅被视为一个建筑元素，而不是一个反映古中国南方干栏民居中席居这一传统生活习惯所形成的空间整体。火塘文化中展示的内容使参观者能够直观地体验土家族人使用火塘空间的方式，因为所有的建筑元素、器皿和器具都是为了适应火塘习俗的生活方式而制作的。

This part of the exhibition creates an impressive spatial experience through the reconstruction of a fire-pit room. In similar exhibitions in the past, the fire-pit was regarded as an architectural element rather than a holistic living space that reflected the traditional lifestyle habitus based on floor-seating of ancient Ganlan dwellings in southern China. In this exhibit, visitors can visualise the way in which people used the fire-pit space because all architectural elements, utensils, and implements are made to adapt to lifestyle based on fire-pit-related customs.

在合宜生境主题中，照片和模型勾勒出土家聚落的自然生态环境与人居生境营造的图景，展示了武陵山区丰富的自然人文景观要素。稻作文化、乡村生活、传统民艺、音乐舞蹈、节庆仪式等是展览中乡村旅游吸引力的重要组成部分。

In the theme of suitable habitat, photos and models outline the natural ecological environment and human habitat creation of the Tujia settlement, showcasing the rich natural and cultural landscape elements of the Wuling Mountain area. Rice cultivation culture, rural life, traditional folk arts, music and dance, festival ceremonies, and other elements are important components of the exhibitions appeal for rural tourism.

6.2 亚洲干栏木构建筑遗产的保护与再生学术研讨会

International Symposium on Conservation and Regeneration of Asian Ganlan Wooden Architectural Heritage

● 研讨会背景及议程

Background and agenda of symposium

本次学术研讨会和展览在比较文化的视野中提供了一种关于建筑遗产保护和再生的全新视野与路径。本次活动所呈现的系列高质量研究与实践成果进一步展示了亚洲干栏木构建筑遗产价值，探索了其保护与再生的策略，为传统聚落的保护与发展提供了重要的样本与典范，也将中国土家族传统建筑智慧的现代化方案、东方优秀传统文化与乡村旅游、乡村振兴融合发展的活态样本呈现给世界。

This academic seminar and exhibition provided a new vision and path for the protection and regeneration of architectural heritage from the perspective of comparative culture. The series of high-quality research and practice achievements presented in this activity further demonstrated the value of the Asian Ganlan wooden architecture heritage, explored the strategies for its protection and regeneration, and provided important samples and models for the protection and development of traditional settlements. It also presented to the world the modernization plan of the traditional construction wisdom of the Tujia nationality in China, and the living samples of the integration and development of excellent oriental traditional culture, rural tourism and rural revitalization.

亚洲干欄木构建筑遗产的保护与再生学术研讨会

International Symposium on Conservation and Regeneration of Asian Ganlan Wooden Architectural Heritage

2022年8月10日
周三 9:00–18:00
9:00-18:00 Wed 10th August 2022

湖北省恩施土家族苗族自治州宣恩县
中意国际建筑研学营
Sino-Italian International Architecture Study Camp, Xuan'en County, Enshi Tujia and Miao Autonomous Prefecture, Hubei Province

联合主办
- 联合国教科文组织 国际文物保护与修复研究中心–亚洲遗产管理学会
- 东南大学中华民族视觉形象研究基地
- 宣恩县人民政府
- 雅伦格文化艺术基金会

承办单位
- 湖北彭家寨旅游开发有限公司
- 东南大学建筑学院

协办单位
- 华中科技大学建筑与城市规划学院
- 重庆大学建筑城规学院
- 威尼斯建筑大学
- 城市与建筑遗产保护教育部重点实验室（东南大学）

媒体支持
- 《建筑学报》
- 《建筑遗产》
- 《Built Heritage》
- 《新建筑》

Co-hosts
- UNESCO ICCROM - Asian Academy for Heritage Management
- Visual Image Research Base of Chinese Nation in Southeast University
- Xuan'en County People's Government
- Fondazione EMGdotART

Co-organizers
- Hubei Pengjiazhai Tourism Development Co. LTD
- School of Architecture, Southeast University

Joint organizers
- School of Architecture and Urban Planning, Huazhong University of Science and Technology
- School of Architecture and Urban Planning, Chongqing University
- Università Iuav di Venezia
- Key Laboratory of Urban and Architectural Heritage Conservation (Southeast University), Ministry of Education, China

Media Support
- *Architectural Journal*
- *Heritage Architecture*
- *Built Heritage*
- *New Architecture*

亚洲干栏木构建筑遗产的保护与再生学术研讨会

2022年8月10日湖北省恩施土家族苗族自治州宣恩县

会议议程

International Symposium on Conservation and Regeneration of Asian Ganlan Wooden Architectural Heritage

August 10, 2022, Xuan'en County, Enshi Tujia and Miao Autonomous Prefecture, Hubei Province

FORUM AGENDA

Session	Time	Content
开幕致辞 Opening Speech		**主持人 Moderator: 董卫教授 Professor DONG Wei**
	09:00–09:10	**夏泽翰 Shahbaz KHAN** 联合国教科文组织驻北京办事处主任，联合国教科文组织驻中华人民共和国、朝鲜民主主义人民共和国、日本、蒙古和韩国代表（线上） Director and Representative of UNESCO Office Beijing, UNESCO Representative to the Peoples Republic's of China, the Democratic People's Republicof Korea, Japan, Mongolia and the Republic of Korea (Online)
	09:10–09:20	**张彤 ZHANG Tong** 东南大学中华民族视觉形象研究基地常务副主任，东南大学建筑学院院长 Executive Deputy Director of Southeast University Chinese National Visual Image Research Base, Dean of School of Architecture, Southeast University
	09:20–09:30	**马里诺·福林 Marino FOLIN** 雅伦格文化艺术基金会主席,威尼斯建筑大学前校长(线上) Chairman of Fondazione EMGdotART, former Rector of Università Iuav di Venezia (Online)
	09:30–09:40	**习覃 XI Qin** 宣恩县委书记 Secretary of Xuan'en County Committee
主旨报告（上午） Keynote speeches (Morning)		**主持人 Moderator: 李保峰教授 Professor LI Baofen**
	09:40–10:10	**常青 CHANG Qing** 中国科学院院士，同济大学建筑与城市规划学院教授 Academician of Chinese Academy of Sciences, Professor of School of Architecture and Urban Planning, Tongji University **报告题目：历史建成环境再生的理念与途径** **Speech Title: Concept and Approach of Historic Built Environment Regeneration**
	10:10–10:30	**理查德·A. 恩格尔哈特 Richard A. ENGELHARDT** 联合国教科文组织亚太地区前任文化事务专员，东南大学客座教授（线上） Former UNESCO Regional Advisor for Culture in Asia and the Pacific, Guest Professor of Southeast University(Online) **报告题目：从亚洲本土遗产保护汲取可持续发展经验** **Speech Title: Learning Lessons in Sustainability from Asian Vernacular Heritage Conservation**
	10:30–10:40	茶歇 Tea Break
	10:40–11:10	**董卫 DONG Wei** 东南大学建筑学院教授，联合国教科文组织文化资源管理教席主持人，亚洲遗产管理学会秘书处(AAHM)负责人 Professor of School of Architecture, Southeast University, UNESCO Chair in Cultural Resource Management, Head of UNESCO-ICCROM-Asian Academy for Heritage Management (AAHM) **报告题目：区域移民史背景下的城乡遗产保护与发展** **Speech Title: Integrated Urban-Rural Heritage Conservation and Development in the Context Regional Migration History**
	11:10–11:40	**陈薇 CHEN Wei** 东南大学建筑学院教授 Professor of School of Architecture, Southeast University **报告题目：早熟榫卯的意义** **Speech Title: The Significance of Precocious Mortise and Tenon**
		午餐 Lunch Break 11:40—14:00
主旨报告（下午） Keynote speeches (Afternoon)		**主持人 Moderator: 郭华瑜教授 Professor GUO Huayu**
	14:00–14:30	**张彤 ZHANG Tong** 东南大学建筑学院院长，东南大学中华民族视觉形象研究基地常务副主任，第17届威尼斯国际建筑双年展官方平行展“两河口，一个土家会聚之地的再生”主策展人 Dean of School of Architecture, Southeast University, Executive Deputy Director of Visual Image Research Base of Chinese Nation in Southeast University, Chief Curator of the official Collateral Events of the 17th International Architecture Exhibition-"Lianghekou, a Tujia Village of Re-Living-Together" **报告题目：干栏木构建筑的自适应与拓扑再生，以宣恩两河口老街为例** **Speech Title: Self-adaptability and Topologica Deformation of Ganlan Wooden Architecture, Taking Lianghekou Tujia Village as a Case**
	14:30–15:00	**李晓峰 LI Xiaofeng** 华中科技大学建筑与城市规划学院副院长，教授 Associate Dean and Professor, School of Architecture and Urban Planning, Huazhong University of Science and Technology **报告题目：文野耕耘：武陵干栏研究40年回眸** **Speech Title: Cultivation in Wilderness and Civilization: Review of 40 Years' Research on Ganlan in Wuling Area**

Session	Time	Content
主旨报告（下午） Keynote speeches (Afternoon)	15:00–15:30	**阿尔多·艾莫尼诺 Aldo AYMONINO** 威尼斯建筑大学建筑与艺术系系主任，教授，联合国教科文组织遗产和城市更新教席教授（线上） Director and Professor of the Department of Architecture and Arts, Università Iuav di Venezia, and UNESCO Chair for Heritage and Urban Regeneration (Online) **报告题目：木构建筑：坚持传统以适应当代使用** **Speech Title: Wooden Architecture: Persistence of Tradition Allows Contemporary Uses**
	15:30–16:00	**覃琳 QIN Lin** 重庆大学建筑城规学院教授 Professor of School of Architecture and Urban Planning, Chongqing University **报告题目：符号与真实：土家族生活空间的建造与渐替** **Speech Title: The Symbolic and the Real: The Construction and Gradual Replacement of Tujia Living Spaces**
	16:00–16:30	**李海清 LI Haiqing** 东南大学建筑学院教授 Professor of School of Architecture, Southeast University **报告题目：效率何为：传统木构在抗战大后方建筑设计中的化用之思** **Speech Title: On Architectural Efficiency: Thinking about the Design Integration of Traditional Timber in the China's Rear Area During World War II**
	16:30–17:00	**拉芮万·奥兰拉特曼尼 Rawiwan ORANRATMANEE** 泰国清迈大学建筑学院院长，教授（线上） Professor, Dean of Faculty of Architecture, Chiang Mai University (Online) **报告题目：地上生活：干栏建筑及其对亚洲文化遗产的贡献** **Speech Title: Living above Ground: Pile Dwellings and Its Contribution to Asian Cultural Heritage**
圆桌讨论	17:00–18:00	**黄居正教授（主持）+ 与会演讲学者** Prof. HUANG Juzheng (Moderator) + Speakers

● 研讨会主旨报告嘉宾发言
Keynote speakers in the symposium

同济大学常青院士进行主旨报告
Keynote speaker Academician Chang Qing from Tongji University makes the presentation

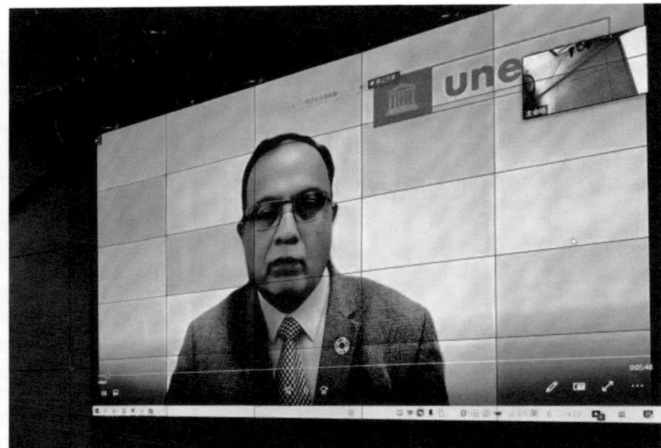

联合国教科文组织驻北京办事处夏泽翰主任进行主旨报告
Keynote speaker Director Shahbaz Khan from UNESCO Office Beijing makes the presentation

东南大学董卫教授进行主旨报告
Keynote speaker Professor Dong Wei from Southeast University makes the presentation

东南大学张彤教授进行主旨报告
Keynote speaker Professor Zhang Tong from Southeast University makes the presentation

东南大学陈薇教授进行主旨报告
Keynote speaker Professor Chen Wei from Southeast University makes the presentation

华中科技大学李保峰教授进行主旨报告
Keynote speaker Professor Li Baofeng from Huazhong University of Science and Technology makes the presentation

华中科技大学李晓峰教授进行主旨报告
Keynote speaker Professor Li Xiaofeng from Huazhong University of Science and Technology makes the presentation

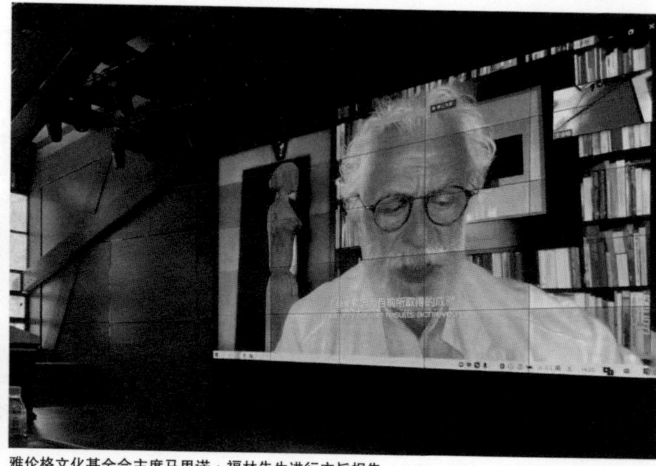

雅伦格文化基金会主席马里诺・福林先生进行主旨报告
Keynote speaker Mr. Marino Folin, Chairman of Fondazione EMGdotART, makes the presentation

参加学术研讨会的现场嘉宾合影（从左至右：宫聪、汤诗旷、董卫、郭华瑜、普拉西多·冈萨雷斯、黄居正、张彤、习覃、常青、邵爱华、李保峰、陈薇、李晓峰、李海清、涂先绪）

Group photography of on-site guests attending the academic seminar (from left to right: Gong Cong, Tang Shikuang, Dong Wei, Guo Huayu, Placido Gonzalez, Huang Juzheng, Zhang Tong, Xi Qin, Chang Qing, Shao Aihua, Li Baofeng, Chen Wei, Li Xiaofeng, Li Haiqing, and Tu Xianxu)

第七章 后记与致谢

CHAPTER 7 POSTCRIPT AND ACKNOWLEGMENTS

张彤
Zhang Tong

教授，博导，
东南大学建筑学院院长
Professor, PhD Supervisor,
Dean, School of Architecture, SEU

这是一次主要由四所大学联合组织和执行的、历时四年的超级教学活动。2018 年 11 月，意大利雅伦格文化基金会主席、威尼斯建筑大学原校长马里诺・福林教授与本次联合教学单位东南大学、华中科技大学的主要发起人来到鄂西武陵山深处的龙潭河谷，确定将古盐道上的商贸古村两河口作为课题载体，组织“两河口，土家盐道古村的再生”国际四校联合研学营。2019 年 4 月底至 5 月初，东南大学、华中科技大学、重庆大学、威尼斯建筑大学、湖北美术学院等 5 所高校的师生来到两河口村开展田野调查和现场测绘，随后师生们回到各自学校进行历史村落的保护与活化更新设计。6 月，研学营师生再次相聚威尼斯，进行设计成果的交流，举办小型展览。之后，威尼斯建筑大学的部分同学以此为基础深化为学位设计，研学营主要老师则策划将成果申请参加第 17 届威尼斯国际建筑双年展，令人欣喜的是申请成功获批策展本届双年展官方平行展“两河口，一个土家会聚之地的再生”。因为疫情，2020 年的双年展推迟到 2021 年，中方师生还是不能去到威尼斯现场，双方在两河口村和威尼斯禅宫举办了国际在线开幕式。在两河口，出外的村民被邀请回来，师生们与当地政府领导、土家村民举办长桌宴共同欢聚庆贺，开幕式得到广泛传播和报道，也成为参加活动各方人员的美好记忆。虽然时处疫情，在威尼斯的展览还是吸引了超过 6 000 人次的现场参观，被双年展组委会评为最受欢迎的 5 个展览之一。展览结束后，展品运回两河口村，被重新组织扩充成为在地常设展“武陵干栏：亚洲木构建筑的文化结晶与样本遗存”，持续向当地居民和游客展示土家建构文化的世界性价值。展览开幕当天，东南大学与联合国教科文组织亚洲遗产管理学会共同举办题为“亚洲干栏木构建筑遗产的保护与再生学术研讨会”，在比较文化视野中探讨建筑遗产保护和再生的全新视野与独特路径。历时四年的教学活动，东南大学、华中科技大学、重庆大学、威尼斯建筑大学、湖北美术学院等 5 所高校的 11 位老师、63 位同学参与其中，从山村现场的田野测绘到最高规格国际学术平台上的展示和交流，完全打破了课堂的边界，师生们在纵横跨越的文化时空中研讨、体验、交流，成为跨文化语境中国际化人才培养的成功案例。

这是一次与教学平行开展的研究、实践和国际交流活动。联合工作营选定的两河口村曾经是湘鄂川黔交界处“川盐济楚”古盐道上的商贸村镇，典型反映了鄂西地区干栏木构建筑的建构特征和商贸市村的聚落形态。近 20 年中，由于人口流失，商贸凋敝，曾经兴盛的古村陷于衰败。针对两河口村与周边土家村寨的乡土建筑遗产调研，使我们能够在国际比较视野中观察、探讨一种建筑遗产保护的亚洲立场。区别于以西方砖石建筑定义的独一性和原真性，干栏木构建筑灵活可变的建构体系使其不拘囿于一时一地的静态样式，具有一种自适应和拓扑性的活态更新机制。这是本次教学活动开启的富有价值的研究视角，相关成果发表在 *Frontiers of Architectural Research*、《建筑遗产》等期刊，我本人在两次威尼斯国际建筑双年展的重要论坛上都对此做了专门的阐述。针对两河口村的联合教学还使得我们的师生面对一个文化遗产保护与利用的世界性问题：如何在当今的旅游业发展中，避免浅表性的消费，既能利用旅游经济带来的发展契机，同时保存和延续历史遗产的真实价值与文化认同；回到我们的专业，如何在建筑学框架内寻求平衡两者的学科韧性。带着对这些问题的认识，中意两国的师生们为两河口古村的保护及龙潭河谷土家村寨的发展提出了完整的规划设计方案，得到宣恩县政府及主管部门的高度认可。

这是一次大学、政府和国际文化机构跨界合作的成功探索。国际联合工作营的教学及相关展览与学术论坛，得到了湖北省恩施土家族苗族自治州、宣恩县、沙道沟镇各级政府和文旅机构的大力支持，还得到了意大利雅伦格文化基金会的全程资助，在此表达衷心感谢！正是他们的支持，使得本次联合工作营成为政产研学联合和国际文化交流的典范案例，产生出丰硕的成果，并不断后续延伸，相关论文在中外高水平期刊上陆续发表；2023 年，两河口民居活态更新再次受邀参加第 18 届威尼斯国际建筑双年展中国国家馆展“Renewal: a Symbiotic Narrative”；师生们的成果汇编成为专著 *Lianghekou, a Tujia Village of Re-Living-Together*，在意大利在线出版。

最后，向本次国际四校联合研学营的发起者、坚定的支持者和执行者、第 17 届威尼斯国际建筑双年展官方平行展策展人之一、威尼斯建筑大学副校长恩里科・丰塔纳里教授致以最真挚和崇高的敬意！他在 2020 年 7 月突然离世，是工作营全体师生的沉痛损失，也是不限于五校的国际建筑教育界的沉痛损失。丰塔纳里教授开阔的学术视野、精湛的专业造诣、如沐春风般的教导指引使我们每一个人深受惠泽。本书的付梓，既是对这次令人难忘教学历程的回顾和记录，也是对恩里科・丰塔纳里教授的深切纪念。

This is a four-year mega teaching activity mainly organized and executed by four universities. In November 2018, Professor Marino Folin, Chairman of Fondazione EMGdotART and former President of IUAV, and the main initiators of the joint teaching activity—SEU and HUST—came to Longtan Valley in the depths of Wuling Mountains in western Hubei. We decided to use Lianghekou, an ancient commercial village on the ancient Salt Road, as the site, and organized a four universities international joint studio on "Lianghekou: Regeneration of an Ancient Tujia Village along the Salt Road". From the end of April to the beginning of May 2019, teachers and students from five universities, including SEU, HUST, CQU, IUAV, and Academy Hubei Institue of Fine Arts (HIFA), came to Lianghekou Village to conduct field investigation and on-site surveys and designed protection and revitalization proposals for this historical village. In June, the teachers and students of the studio gathered in Venice again to exchange design achievements and hold a small exhibition. Afterwards, some IUAV students took this as a basis for their master's degree theses. Furthermore, key teachers of the studio enhanced design achievements and applied to participate in the 17th International Architecture Exhibition—La Biennale di Venezia. It is gratifying that the application was successfully approved as this biennale's official collateral exhibition "Lianghekou, a Tujia Village of Re-Living-Together". Due to the pademic, the 2020 Biennale was postponed to 2021. As Chinese teachers and students were still unable to go to Venice in 2021, we held an international online opening ceremony paralleled in Lianghekou Village and Venice's Palazzo Zen. In Lianghekou, the villagers who had gone out were invited back. The teachers and students held a long table banquet with local government leaders and Tujia villagers to celebrate together. The opening ceremony was widely spread and reported, and it became a good memory of all parties involved in the event. Despite the epidemic, the exhibition in Venice attracted more than 6,000 visitors, and was rated as one of the five most popular exhibitions by the Biennale's Organizing Committee. After the exhibition, the exhibits were transported back to Lianghekou Village, where they were reorganized and expanded into a local permanent exhibition "WULING Ganlan: Splendid Archi-tectonic Legacy of Asian Wooden Architecture". It consistently demonstrates the global value of Tujia construction culture to local residents and tourists. On the opening day of the exhibition, SEU and UNESCO Asian Academy for Heritage Management co-organized the "International Symposium on Conservation and Regeneration of Asian Ganlan Wooden Architectural Heritage" to explore a new vision and unique path of architectural heritage protection and regeneration from a comparative perspective. During the four-year teaching activities, 11 teachers and 63 students from five universities, including SEU, HUST, CQU, IUAV, and HIFA, participated, from field surveying and mapping in the mountain village to the highest standard international academic platform. The display and communication along this journey completely broke the boundaries of the classroom. Teachers and students discussed, experienced and communicated across the borders of culture, time and space, becoming a successful case of international talent cultivation in a cross-cultural context.

This is a research, practice and international exchange activity carried out in parallel with teaching. The jointly selected Lianghekou Village was once a commercial village on the ancient Salt Road of "*Chuanyan Ji Chu*" at the junction of Hunan, Hubei, Sichuan and Guizhou Provinces. In the past 20 years, due to the loss of population and the decline of commerce and trade, this once-prosperous ancient village fell into decline. The research on the vernacular architectural heritage of Lianghekou Village and surrounding Tujia villages enables us to observe and explore an Asian standpoint of architectural heritage protection from an international comparative perspective. Different from the uniqueness and authenticity defined by Western masonry-dominated architecture, the flexible and changeable construction system of Ganlan wooden architecture makes it not limited to the static style of a certain time and location, but has an adaptive and topological dynamic mechanism. This is a valuable research perspective that is opened up by this teaching activity. Relevant research achievements have been published in various academic journals including *Frontiers of Architectural Research* and *Heritage Architecture*. I have made special presentations on this topic in two significant forums of the Venice Architecture Biennale. The joint teaching of Lianghekou Village also makes our teachers and students face a worldwide problem of cultural heritage protection and utilization: how to avoid superficial consumption in today's development of tourism while making use of the tourism economy to bring opportunities for progress and preserving and continuing the real value and cultural identity of heritage; for our profession, how to balance these two aims within the framework of architecture as a resilient discipline. With an understanding of this issue, teachers and students from China and Italy put forward a complete planning and design proposal for the protection of Lianghekou Village and the development of Tujia villages along Longtan Valley. This proposal was highly recognized by the Xuan'en County government and other managing authorities.

This is a successful exploration of cross-border cooperation among universities, governments and international cultural institutions. The teaching and related exhibitions and academic forums of the studio have received strong support from governments at all levels in Enshi Tujia and Miao Autonomous Prefecture, Xuan'en County, and Shadaogou Town, as well as their related cultural and tourism institutions. Fondazione EMGdotART has given full funding. I would like to express my heartfelt thanks for the full support! It is their support that makes this joint teaching studio a model case of government-industry-research cooperation and international cultural exchange, which has produced fruitful outcomes and has been continuously extended: related papers have been published in high-level journals at home and abroad; the living renewal of Lianghekou residents was once again invited to participate in the Chinese pavilion' exhibition "Renewal: a Symbiotic Narrative" at the 18th Venice International Architecture Biennale; the achievements of teachers and students were compiled into a monograph *Lianghekou, a Tujia Village of Re-Living-Together*, published online in Italy.

Finally, I would like to extend my most sincere and highest respect to Professor Enrico Fontanari, the initiator, staunch supporter and executor of this four universities international joint studio, one of the curators of the official collateral exhibition of the 17th International Architecture Exhibition—La Biennale di Venezia, and the vice President of IUAV. His sudden death in July 2020 was a painful loss for all the teachers and students in the joint teaching studio, and it was also a painful loss for the international architectural education community not limited to the five universities. Professor Fontanari's broad academic vision, superb professional attainments, and spring-like teaching and guidance have greatly benefited each of us. The publication of this book is not only a review and record of this unforgettable teaching process, but also a deep memory of Professor Enrico Fontanari.

马里诺・福林
Marino Folin

雅伦格文化艺术基金会主席
Chairman of Fondazione EMGdotART

近年来，中国进行了一系列的乡村振兴项目，旨在保护远离大城市的农村和小镇的物质、经济和社会发展，改善其生活条件。与西方相比，这些项目无论在规模、持续性和干预方式上都是空前的。几乎所有的振兴项目在古代都是聚居地或者农牧经济中心，且往往是少数民族居住区。在这些地方，少数民族仍以植根于他们的语言、文化和物质传统的生活方式生活着。

目前进行的乡村振兴项目中，最新、最引人注目的项目坐落于湖北省宣恩县龙潭河西岸、武陵山坡上。这里自古就是中国少数民族土家族聚居地。现在，他们的村庄被列为中国少数民族特色村寨之一，也是国保单位。

为了振兴彭家寨，宣恩县政府委托华中科技大学建筑与城市规划学院的李保峰教授，负责彭家寨泛博物馆项目的规划和建筑设计。这次乡村振兴项目以及所有与之相关的活动，都离不开华中科技大学建筑与城市规划学院的支持和学院学生的积极参与。土家泛博物馆项目涵盖范围相对较广，涉及对从两河口村到白果坝发电厂的龙潭河山谷以及树木葱郁的武陵山坡的保护、再生和发展，三方面密不可分。山谷中，蜿蜒着千年川盐古道，沿途更有众多的土家族和苗族村落。土地规划的主要目标是保护文化、建筑和景观遗产，对符合本地气候、土壤成分和地域形态制约的传统农业和手工艺进行恢复，并发展本地的可持续的环保的旅游业。

考虑到该项目在国际上的重要性，在 2019 年第 17 届威尼斯国际建筑双年展之际，雅伦格文化艺术基金会推动了“活化——以中国土家泛博物馆为例”项目在威尼斯禅宫的展示。首次展示后，雅伦格文化艺术基金会又推动了威尼斯建筑大学与位于南京、负责两河口再生计划的东南大学之间的合作协议（协议的基础为雅伦格文化艺术基金会奖，该奖项由雅伦格文化艺术基金会发起）。在两校师生的努力和地方政府的支持下，两校联同华中科技大学和重庆大学共同为两河口的再生制定城市规划和建设项目。

2019—2021 年，东南大学、威尼斯建筑大学、华中科技大学和重庆大学的师生与宣恩县合作完成的项目“两河口：一个土家会聚之地的再生”，于第 17 届威尼斯国际建筑双年展展出（2021 年 5 月—11 月）。展览在宣恩县政府、东南大学、威尼斯建筑大学、华中科技大学、重庆大学、未来城市产业创新中心和未来城市与数字建筑协会的支持下，由雅伦格文化艺术基金会举办。

彭家寨和两河口的乡村振兴项目具有示范效应。在西方国家，有数以千计的历史村落和小型城市中心处在衰落或废弃中，主要分布在偏远的丘陵或山区中。在物质意义上对历史中心进行重建方面，西方国家有一些很好的实践案例。如不加以修复，这些地区无疑会逐渐衰落、被人遗忘。但西方国家的重建几乎总伴随着中产阶级化进程，破坏了当地原有的社会体系，彻底抛弃传统的经济和生产方式。彭家寨和两河口乡村振兴项目的示范意义在于，区位的振兴与保存发展现有的经济和社会体系、传承发扬物质和非物质文化遗产齐头并进。

“两河口，一个土家会聚之地的再生”展览，向国际观众展示了两河口乡村振兴项目的示范意义。

In recent years China has started and implemented a number of rural regeneration programs unequaled in the West, by size, continuity and methods of intervention, aimed at the physical, economic and social protection and the improvement of living conditions of villages and small towns of the country far from the big urban centers. In almost all cases, they are centers of ancient settlements and of agricultural-pastoral economy, very often inhabited by ethnic minorities, in which still survive the ways of life rooted in their linguistic, cultural and material traditions.

Among the most recent and exciting rural regeneration projects in progress is that of villages located along the west bank of the Longtan River, on the slopes of Mount Wuling (Hubei Province, Xuan'en County) in a valley inhabited since ancient times the Tujia Nationality, one of the ethnic minorities of China. These villages are now included in the list of outstanding minority villages of China, and listed as national key protected units.

As part of the actions aimed at regenerating the village of Pengjia Village located in the valley, the Xuan'en County has entrusted the task both of the territorial plan and of the project of main buildings, among which the Pan-museum of the Tujia minority, to Prof. Li Baofeng of the School of Architecture and Urban Planning (Huazhong University of Science and Technology, HUST). The projects, with all the activities connected to them, have been developed with the support of the School of Architecture and Urban Planning of HUST, and with the active participation of the students of the School. The Tujia Pan-museum project is part of a more general territorial plan, of preservation, regeneration and development—the three aspects are inextricably linked to each other—concerning the Longtan River valley, from the village of Lianghekou to the power plant of Baiguoba, including the immediate wooded slopes of the Wuling mountains. The valley is crossed by the millennial Sichuan's Salt Road, along which there are numerous villages of Tujia and Miao minorities. The main goals of the territorial plan are the protection of cultural, architectural and landscape heritage; the enhancement of traditional agricultural and craft activities, in full compliance with environmental conditions dictated by the climate, the composition of the soil, the morphology of the sites; the promotion of a local, sustainable and conscious tourism.

Considering the great importance, at the international level, of these projects, the Fondazione EMGdotART promoted the exhibition of the "Revitalization—Taking the Chinese Tujia Pan-museum as an Example" project, at Palazzo Zen in Venice, on the occasion of the 17th International Architecture Exhibition—La Biennale of Venice 2019.

Following this first presentation, the Foundation later promoted a cooperation agreement based on the Fondazione EMGdotART Award (initiated by Fondazione EMGdotART supporting the academic exchange between SEU and IUAV) between the Southeast University, Nanjing, in charge of the regeneration plan of the village of Lianghekou, and the Università Iuav di Venezia, in order to draw up urban planning and building projects for the regeneration of Lianghekou, with the joint participation of teachers and students of both the universities and the support of local authorities. The initiative was joined by Huazhong University of Technology and Chongqing University.

The projects produced in 2019-2021 by professors and students of Southeast University, Università Iuav di Venezia, Huazhong University of Science and Technology, Chongqing University, with the collaboration of Xuan'en County are now presented at the exhibition "Lianghekou, a Tujia Village of Re-Living-Together", a collateral event of the 17th International Architecture Exhibition—La Biennale di Venezia (May—November 2021). The exhibition has been organized by Fondazione EMGdotART with the support of the municipality of Xuan'en County, Southeast University, Università Iuav di Venezia, Huazhong University of Science and Technology, Chongqing University, Future Urban Industrial Innovation Center, and Future City and Digital Architecture Association.

The projects for the regeneration of Pengjia Village and Lianghekou, are exemplary projects. In Western countries, there are thousands of historical villages and small urban centers in decline or abandoned, located mainly in peripheral hill or mountain areas. In Western countries there are good examples of physical restoration and rehabilitation of historical centers, otherwise destined to progressive degradation and abandonment, but almost always this rehabilitation took place together with processes of gentrification, of destruction of the previously existing social system, and of total abandonment of traditional economies and productions. The projects for Pengjia Village and Lianghekou are exemplary because, in these cases, urban and territorial regeneration goes hand in hand with the preservation and development of the existing economic and social system and with the enhancement of the material and immaterial culture inherited from the past.

The exhibition "Lianghekou, a Tujia Village of Re-Living-Together" presented to an international audience the regeneration project of Lianghekou as an exemplary model to follow.

阿尔多·艾莫尼诺
Aldo Aymonino

威尼斯建筑大学建筑与艺术系系主任
联合国教科文组织遗产与城市更新教席教授
Director of Department of Architecture and Arts，IUAV University of Venice
UNESCO Chair for Heritage and Urban Regeneration

“两河口：土家盐道古村的再生” 国际四校联合研学营是一次发现有趣的中国文化的现实之旅，提供给参与者发掘一个远离传统的旅游和国际商务发展路径的机会。能够在中国最古老的少数民族聚居区之一，也是最核心、最具特色的地区之一进行现场调研，这样独特的机深刻地影响了学生和老师处理项目的方式。首先，很显然，对古盐道的任何干预都涉及对现有建筑物尺寸、材料和空间的处理，同时也涉及现代语言和文化遗产之间微妙的关系。尽管我们已决定，不试图对过去几十年中丢失或被剧烈改变的部分进行风格重建，但依然必须面对和衡量具有如此强烈暗示意味的风格形式，设计结果也因此受到了直接的制约和影响。意大利学生虽然来自与调研现场截然不同的社会文化背景，但在所有项目中均尽力面对当地现实，采取与“场所精神”相协调的设计策略。

此外，项目考虑到两河口村有可能成为整个土家族盐道区域文化综合体的门户，这个综合体涵盖由11个村庄组成的、长达约10千米的范围。

由于被列为国保单位，土家族村落（彭家寨）会成为新旅游景区项目的一部分。

目前的地域规划和设计综合考量了新道路基础设施以及土地消耗量的限制措施，对历史、文化、建筑遗产以及更适合本地的景观的保护等需求和条件，从而推动当地发展可持续的、符合紧急环境保护需要的、能促进当地社区发展的旅游业，并对一定程度上造成并加剧本地人口减少的国内人口流动现状进行扭转。

The four universities international joint studio, "Lianghekou, Regeneration of an Ancient Tujia Village along the Salt Road" was an opportunity to discover an interesting local Chinese cultural reality, far from the conventional routes of tourism and international business. The unique possibility of being able to work on site in the places of one of the most ancient cultural minorities of China and in one of its most integral and characteristic landscapes has deeply influenced the behavior with which both students and teachers have approached the project. First, it was evident that interventions on the ancient Salt Road would have to deal with the caliber, materials and spaces of the pre-existing buildings. Another aspect concerned the delicate relationship between contemporary language and heritage. Although we were convinced that we did not want to attempt the stylistic reconstruction of the parts missing or deeply altered in the past decades, the need to confront and measure up against such a strongly connoted stylistic-formal integrity immediately conditioned and seduced the design results. All the Italian students, although coming from socio-cultural contexts so different from the model under examination, have tried to confront themselves with the local reality, using design strategies in tune with the "genius loci".

Moreover, the possibility has been taken into account that the village of Lianghekou may become the gateway to the entire cultural complex of the "Tujia Salt Road Valley", which includes a territorial system of eleven villages for a length of about ten kilometers.

The Tujia villages are part of the program of identification of possible new tourist attractions by virtue of their inclusion in the lists of national heritage sites.

The new territorial scenario that is taking shape imposes a careful reflection on the modalities and the measures to limit the consumption of land that the opening of the new road infrastructure will induce, on the preservation of historical, cultural, architectural heritage as well as of the landscape matrixes more proper of the places, on the addresses for the development of sustainable tourism compatible with the needs of safeguarding of environmental emergencies and with the growth of local communities, also able to reverse the more or less recent internal migration flows that have induced and encouraged the current conditions of depopulation of the places.

※ 除特殊注明外，本书中所有图片由本次国际联合教学工作营提供，包括调研照片、图绘表格、模型照片等由四校师生拍摄、绘制。

Unless otherwise specified, all images in this book are provided by the international joint teaching workshop, including research photos, drawing tables, model photos, etc., which were taken and drawn by teachers and students from the four universities.